NIST Special Publication 260-143

Standard Reference Materials®

Certification of the Rheological Behavior of SRM 2490, Polyisobutylene Dissolved in 2,6,10,14-Tetramethylpentadecane

Carl R. Schultheisz[1] and Stefan D. Leigh[2]

[1] Polymers Division, Materials Science and Engineering Laboratory
[2] Statistical Engineering Division, Information Technology Laboratory
National Institute of Standards and Technology
Gaithersburg, MD 20899-8544

U.S. DEPARTMENT OF COMMERCE, *Donald L. Evans, Secretary*
TECHNOLOGY ADMINISTRATION, *Phillip J. Bond, Under Secretary of Commerce for Technology*
NATIONAL INSTITUTE OF STANDARDS AND TECHNOLOGY, *Arden L. Bement, Jr., Director*

Issued February 2002

Table of Contents

Abstract	1
1 Introduction	2
2 Preparation, Bottling, and Sampling of SRM 2490	3
3 Testing	4
3.1 Homogeneity Testing	4
3.2 Steady Shear Testing for Certification	5
3.2.1 Testing at 25 °C	6
3.2.2 Testing at 50 °C	7
3.2.3 Testing at 0 °C	9
3.3 Dynamic Testing	10
4 Analysis of Sources of Uncertainty	12
4.1 Steady Shear Testing in the Cone and Plate Fixtures	12
4.1.1 Uncertainty in the Independent Variables: Temperature and Shear Rate	13
4.1.2 Uncertainties in the Viscosity and First Normal Stress Difference	14
4.1.2.1 Uncertainty η and N_1 Arising from Uncertainty in Temperature	17
4.1.2.2 Uncertainty in η and N_1 Arising from Uncertainty in Transducer Readings	18
4.1.2.3 Uncertainty in η and N_1 Arising from Uncertainty in Rotation Rate	18
4.1.2.4 Uncertainty in η and N_1 Arising from Uncertainties Associated with Geometry	18
4.1.2.4.1 Gap	18
4.1.2.4.2 Cone Angle	19
4.1.2.4.3 Cone/Plate Diameter	19
4.1.2.4.4 Cone Truncation	20
4.1.2.4.5 Tilt	20
4.1.2.4.6 Concentricity	20
4.1.2.5 Uncertainty Associated with Solvent Evaporation	21
4.1.2.6 Effects of Shear Heating	21
4.1.2.7 Inertia-Driven Secondary Flows	21
4.1.2.8 Edge Effects	22
4.1.2.9 Surface Tension	22
4.2 Dynamic Testing in Parallel Plates	32
4.2.1 Uncertainty in the Independent Variables: Temperature and Frequency of Oscillation	32
4.2.2 Uncertainties in G' and G"	32
4.2.2.1 Uncertainties Associated with Temperature	36
4.2.2.2 Uncertainties Associated with Frequency of Oscillation	36

	4.2.2.3 Uncertainties Associated with Cross-correlation of the Transducer Signal and Oscillation Magnitude	37
	4.2.2.4 Uncertainties Associated with Geometry	38
	4.2.2.4.1 Gap	38
	4.2.2.4.2 Plate Diameter	38
	4.2.2.4.3 Tilt	39
	4.2.2.4.4 Concentricity	39
	4.2.2.5 Solvent Evaporation	39
	4.2.2.6 Inertia	39
	4.2.2.7 Edge Effects	39
5 Conclusion		64
6 References		74

List of Tables

Table 1.	Uncertainty in the Shear Rate	14
Table 2.	Parameters for $\eta(\dot{\gamma}, T)$, $a(T)$ and $N_1(\dot{\gamma}, T)$ in the models found in equations (6) through (8)	15
Table 3.	Components of Standard Uncertainty of the Viscosity η	23-28
Table 4.	Components of Uncertainty in the First Normal Stress Difference N_1	29-31
Table 5.	Parameters for G' (Ω, T), G'' (Ω, T) and $a(T)$ in the models found in equations (39) and (40)	35
Table 6.	Components of Standard Uncertainty in the Storage Modulus G'	40-51
Table 7.	Components of Standard Uncertainty in the Loss Modulus G''	52-63
Table 8.	Viscosity and First Normal Stress Difference (N_1) with Combined Standard Uncertainties and Models Fit to the Data	65-67
Table 9.	Storage Modulus G' and Loss Modulus G'' with the Combined Standard Uncertainties and Models Fit to the Data	68-73

Certification of the Rheological Behavior of SRM 2490, Polyisobutylene Dissolved in 2,6,10,14-Tetramethylpentadecane

Carl R. Schultheisz[1] and Stefan D. Leigh[2]
[1]Polymers Division
[2]Statistical Engineering Division
National Institute of Standards and Technology
Gaithersburg, MD 20899

Final report prepared for the sponsors, the NIST Standard Reference Materials Program (SRMP).

ABSTRACT

The certification of the rheological properties of Standard Reference Material® (SRM) 2490, a non-Newtonian fluid consisting of polyisobutylene dissolved in 2,6,10,14-tetramethylpentadecane, is described. The viscosity and the first normal stress difference were measured in steady shear at rates between 0.001 s^{-1} and 100 s^{-1} at 0 °C, 25 °C and 50 °C. The linear viscoelastic storage modulus G' and loss modulus G" were also measured in dynamic oscillatory measurements between 0.04 rad/s and 100 rad/s in the temperature range between 0 °C and 50 °C and master curves calculated using time temperature superposition.

1. Introduction

This report describes the certification of the rheological properties of Standard Reference Material® (SRM) 2490, a non-Newtonian fluid consisting of polyisobutylene dissolved in 2,6,10,14-tetramethylpentadecane. This fluid demonstrates shear thinning (decreasing viscosity with increasing shear rate) and development of normal stresses, which are rheological behaviors common to polymeric materials [1-6]. The viscosity and the first normal stress difference were measured in steady shear at shear rates between 0.001 s^{-1} and 100 s^{-1} at 0 °C, 25 °C and 50 °C. The linear viscoelastic storage modulus G' and loss modulus G" were also measured in dynamic oscillatory measurements between 0.04 rad/s and 100 rad/s in the temperature range between 0 °C and 50 °C, and master curves were calculated using time temperature superposition.

SRM 2490 replaces an earlier material, SRM 1490, which consisted of polyisobutylene dissolved in normal hexadecane. SRM 1490 was certified in 1977 [7], but discontinued in 1990 when deviations from the certified properties were identified. Among the possible reasons for the deviation from the certified properties were the packaging of the material and evaporation of the solvent. SRM 1490 was delivered in large (20 liter, 5 gallon) plastic containers, with material transferred to bottles for sale as needed; material taken from different large containers might have led to a difference in properties from one lot of bottles to the next. Similarly, evaporation of the solvent through the plastic containers over time could have led to changes in properties; in the measurements that showed a deviation, the viscosity was found to be higher than the certified value, which is consistent with solvent evaporation. In addition to these problems, the normal hexadecane used as a solvent also limited the useful temperature range of the material, since normal hexadecane crystallizes at 18 °C (64 °F) [8].

A polyisobutylene solution was still considered a good choice as an SRM, because the saturated hydrocarbon structure of polyisobutylene is chemically stable, and polyisobutylene poses no health risks. Also, polyisobutylene is widely available commercially and therefore cost effective. In addition, polyisobutylene solutions have been widely used as model non-Newtonian fluids, because they demonstrate significant shear thinning and relatively large normal stresses [9-13].

In an effort to overcome the difficulties associated with the normal hexadecane used as the solvent for SRM 1490, a new solvent was chosen for SRM 2490. The new solvent (2,6,10,14-tetramethylpentadecane [14]) also has a saturated hydrocarbon structure for chemical stability, but it was chosen to have a branched alkane structure to inhibit crystallization. Differential scanning calorimetry measurements show no evidence of crystallization or vitrification of this solvent down to temperatures as low as −60 °C. The new solvent also has a slightly higher molar mass than hexadecane, so its rate of evaporation is somewhat lower. This material is also commercially available from several sources, and its cost is comparable to that of normal hexadecane.

In an effort to eliminate possible problems of variations between large containers and/or differences arising from bottling at different times, all of the polymer solution was mixed in one large container and then bottled in 100 ml quantities at the same time.

The common name for normal hexadecane is *cetane*, which comes from the Latin word *Cetacea*, the name for the order that contains whales and porpoises [15]. Cetane is found in sperm whale oil. The common name for 2,6,10,14-tetramethylpentadecane is *pristane*, which comes from *pristis*, the Latin word for shark [15]. Pristane is found in shark liver oil [14].

2. Preparation, Bottling, and Sampling of SRM 2490

SRM 2490 consists of polyisobutylene (Aldrich Chemical, catalog number 18,146-3, CAS number 9003-27-4) dissolved in 2,6,10,14-tetramethylpentadecane (Aldrich Chemical, catalog number T2,280-2, CAS number 1921-70-6); the solution contains a mass fraction of 0.114 polyisobutylene. The supplier identifies the polyisobutylene as being stabilized with 500 ppm 2,6-di-*tert*-butyl-4-methylphenol, and gives the average Mv ca. 1,200,000; Mw ca. 1,000,000; Mn ca. 600,000 as measured by GPC/MALLS; the density of the polyisobutylene is given as 0.898 g/cm^3. The density of 2,6,10,14-tetramethylpentadecane is given by the supplier as 0.785 g/cm^3. The specification of a concentration having a mass fraction of 0.114 polyisobutylene was extrapolated from the viscosity of a solution prepared at NIST with a mass fraction of 0.10 polyisobutylene [12] and the concentration dependence of solutions of polyisobutylene dissolved in n-hexadecane [10]. The solution was intended to have a zero-shear-rate viscosity of 200 Pa·s at 25 °C. Subsequently, a solution having a mass fraction of 0.114 polyisobutylene dissolved in 2,6,10,14-tetramethylpentadecane prepared at NIST was found to have a zero-shear-rate viscosity of approximately 150 Pa·s at 25 °C [13].

Mixing and bottling of SRM 2490 was contracted out to Cannon Instrument Company. Quantities of polyisobutylene and 2,6,10,14-tetramethylpentadecane in the correct proportions were placed in four separate containers, and once the dissolution process was complete, the contents of the four containers were mixed together in a single container and then bottled. Initially, it was hoped that the dissolution process could proceed at room temperature with minimum agitation, in order to minimize degradation of the polyisobutylene [11]. The dissolution process was found to be very slow, however, so intermittent heating up to 50 °C and additional mechanical agitation were employed to speed up the process. Some degradation of the polyisobutylene may have occurred during mixing, as the zero-shear-rate viscosity of the resulting solution was approximately 100 Pa·s at 25 °C, as compared to the value of 150 Pa·s for the solution prepared at NIST. Some of that difference might also be attributed to different lots of the polyisobutylene. The solution was packaged in quantities of 100 ml in amber glass bottles with phenolic caps and conical polyethylene cap liners, and 439 bottles were delivered to NIST.

The bottles were numbered and filled sequentially. Ten bottles were taken as samples for homogeneity testing and for certification. One bottle was chosen randomly from each successive sequence of 45 bottles (1-45, 46-90, etc.). This procedure was intended to uncover any systematic variation in concentration that might have occurred in the course of the bottling process.

Before taking a sample from any bottle for testing in the rheometer, the bottle was turned end over end at a rate of 1 revolution per 10 min for a minimum of 30 min. For turning, the bottles were held with a three-jaw clamp attached to a small motor/gear assembly (McMaster -Carr). Turning the bottles was intended to ensure that the material within each bottle was homogeneous, in case there was any settling due to gravity.

3. Testing

All rheological testing was carried out in a Rheometric Scientific, Inc. ARES controlled-strain rheometer. This rheometer employs a force rebalance transducer that measures both torque and normal force. The torque transducer has two ranges, a low range with higher sensitivity for torque with magnitude up to 0.02 N·m (200 g_f·cm, where g_f indicates a unit of gram force, which is not an acceptable unit of the SI, but is commonly used), and a high range for torque with magnitude up to 0.2 N·m (2000 g_f·cm). The normal force transducer has a single range capable of measuring normal forces up to 20 N (2000 g_f) in magnitude. The rheometer transducer was calibrated each day before any testing. Calibration was accomplished following the manufacturer's instructions [16]; the procedure consists of hanging a known mass from a fixture mounted to the transducer to apply a known torque or normal force. The calibration masses were checked on routinely calibrated electronic balances at NIST (OHAUS GA200D for masses of 200 g or less, Mettler PE 3600 for masses greater than 200 g) and found to be within 0.02 % of the specified mass. The dimensions of the calibration fixture were also measured at NIST and found to be within 0.2 % of the specification. For dynamic measurements, the phase angle was also calibrated each day before any testing, again following the manufacturer's instructions.

Five sets of tests were conducted. The first set of tests investigated the homogeneity of the material delivered; the object of these tests was to compare the zero-shear-rate viscosity at 25 °C, which was used as an acceptance criterion. Four sets of certification tests were then performed: steady shear at 25 °C, 50 °C and 0 °C to determine the viscosity and first normal stress difference as functions of the shear rate, and dynamic tests over a range of frequency during temperature sweeps from 0 °C to 50 °C to determine the linear viscoelastic properties.

The steady shear tests were performed using cone and plate fixtures, and the dynamic tests were performed using parallel plate fixtures. The fixtures were cleaned with acetone before each experiment.

Fresh nitrile gloves were worn for cleaning the fixtures and for loading the samples. Gloves were worn both to prevent skin contact with the material and to prevent contaminating the fixtures or the samples with grease or dirt from the operator's hands. The material safety data sheet for 2,6,10,14-tetramethylpentadecane does not indicate any specific health hazards, but does suggest the possibility of skin irritation, as with most organic solvents. No unusual measures were taken for disposing of the samples after testing.

Temperature control in the ARES is maintained by a forced air system into an insulated chamber surrounding the test section; the air can be cooled with liquid nitrogen or heated with electrical resistance coils. The temperature control was calibrated before each series of certification tests using a NIST-calibrated thermistor [17]. The combined standard uncertainty in the temperature was estimated to be 0.1 °C.

3.1 Homogeneity Testing

The criterion for accepting the delivery of the bottled solution was that the zero-shear-rate viscosity measured at 25 °C from a representative sample of the bottles should have a maximum spread of 4 % of the average zero-shear-rate viscosity. Because the time frame for homogeneity testing was limited, and because a smaller amount of data was required than for the steady shear certification tests, some of the parameters for the homogeneity tests were chosen differently than for the certification tests. The fixtures used were 50 mm diameter cone and plate, with a cone angle specified to be 0.0197 rad. The cone is truncated, and the distance between the point where the vertex of the cone should be and the plane of truncation is 0.048 mm. The dimensions of the fixtures were measured at NIST and found to agree with the manufacturer's specification within the uncertainties. The testing was performed at 25 °C, with the temperature control based on the temperature read by platinum resistance thermometers near the forced air inlets (ARES Control Mode 3). The tests were run using the Steady Rate Sweep template supplied with the ARES software, with the tests specified to start at a shear rate of 0.01 s^{-1} and then at increasing rates, taking five points per decade up to a maximum of 100 s^{-1}. All of these tests were performed using the more sensitive low range of the torque transducer; the transient overshoot in the viscosity typically led to an overload in the torque at a shear rate of 16 s^{-1}. Measurements at each shear rate were made in both the clockwise and counterclockwise directions, with the shear applied for 100 s and then measurements of the torque and normal force averaged over the next 30 s. Before loading each sample, the cone and plate fixtures were installed and set to a nominal gap of 1 mm. The temperature control chamber was then closed and the fixtures brought to 25 °C for 30 min. The gap was then zeroed using the Autozero capability of the ARES (the fixtures brought together until they touched, to establish the baseline from which to set the gap to the proper dimension). The temperature control chamber was then opened, and the upper fixture was raised approximately 60 mm. As mentioned above, each bottle was turned end over end for at least 30 min before testing to thoroughly mix the material inside; a sample was then spooned onto the lower plate fixture, and the upper cone fixture was lowered to the specified offset of 0.048 mm. The sample was then trimmed flush with the edges of the fixtures, and the temperature control chamber closed. The specified Steady Rate Sweep test was then started; the test sequence included a 30 min hold at 25 °C to allow the temperature to reach steady state. Three samples were tested from each of the ten bottles; the thirty samples were tested in a random order. Typically, three samples were tested each day.

The testing indicated that the homogeneity of the material was within the specified acceptance criteria. Statistical tests using analysis of variance (ANOVA) showed that test-sequence effects were not statistically significant compared to the variability among bottles.

3.2 Steady Shear Testing for Certification

The steady shear testing for certification was similar to the tests for homogeneity, but there were some changes in the test procedure and parameters. A broader range of shear rates was employed, and the mode of temperature control was changed to use a platinum resistance thermometer that is in contact with the underside of the lower plate fixture (ARES control mode 2). Tests were carried out at 25 °C, 50 °C and 0 °C (in that order), and comparison with a NIST-calibrated thermistor was used to adjust the parameters in a look-up table for the temperature controller before beginning a series of tests at a new temperature.

3.2.1 Testing at 25 °C

The fixtures were again 50 mm diameter, 0.0197 rad cone and plate, with the cone truncated at 0.048 mm from its vertex. The tests were run using the Steady Rate Sweep template supplied with the ARES software, taking five points per decade. Tests were begun using the more sensitive low range of the torque transducer, sweeping from an initial shear rate of 0.001 s^{-1} up to a shear rate of 10 s^{-1}. The instrument was then switched to use the high range of the torque transducer, and a rate sweep from 10 s^{-1} to 100 s^{-1} was performed. (The switch was necessary, because the torque at the higher shear rates was sufficient to overload the low range of the transducer.) Measurements at each shear rate were made in both the clockwise and counterclockwise directions, with the shear applied for 30 s and then measurements of the torque and normal force averaged over the next 30 s. Before loading each sample, the cone and plate fixtures were installed and set to a nominal gap of 1 mm. The temperature control chamber was then closed and the fixtures brought to 25 °C for 30 min. The gap was then zeroed using the Autozero capability of the ARES (the fixtures brought together until they touched, to establish the baseline from which to set the gap to the proper dimension). The temperature control chamber was then opened, and the upper fixture was raised approximately 60 mm. Each bottle was turned end over end for at least 30 min before testing to thoroughly mix the material inside; a sample was then spooned onto the lower plate fixture, and the upper cone fixture was lowered to a position of 0.046 mm. The sample was then trimmed flush with the edges of the fixtures, and the cone was repositioned to the specified gap of 0.048 mm. This procedure was used in an effort to compensate for the expansion of the sample when changing from room temperature to the test temperature of 25 °C, in order to achieve the appropriate spherical sample geometry. The temperature control chamber was closed, and the specified Steady Rate Sweep test was started; the test sequence included a 30 min hold at 25 °C to allow the temperature to reach steady state. Two samples were tested from each of the ten bottles; the twenty samples were tested in a random order.

In reducing the data, the viscosity measurements taken at 10 s^{-1} using the low transducer range and the high transducer range were averaged, although typically the difference was much smaller than the uncertainty of the measurements. Averaging measurements made using rotations in both directions eliminates any problems of an offset in the zero position for the viscosity. The normal force measurement, however, is sensitive to the zero position and is strongly affected by temperature fluctuations that cause transient volumetric changes. The zero position of the normal force measurements was adjusted by averaging the first six measurements and subtracting that value from all subsequent measurements. The normal force in that range (0.001 s^{-1} to 0.01 s^{-1}) is well below the sensitivity of the transducer and so can be taken as zero. In changing the transducer from the low torque range to the high torque range, the difference between the two normal force measurements (taken at the same shear rate of 10 s^{-1}) was then subtracted from the subsequent measurements taken with the transducer using the high torque range.

The total time for a single test was approximately 90 min, including the 30 min delay to ensure thermal steady state before measurements were begun. A possible source of error in the measurement is evaporation of the solvent during the experiment caused by the forced air temperature control system. To investigate whether evaporation of the solvent caused a significant change in the sample over the duration of the test, two experiments were performed using 90 min delays inside the temperature chamber before measurements were begun. These tests indicated that the effects of solvent evaporation are not significant compared to the uncertainties in the measurements for tests lasting 90 min inside the chamber at 25 °C. Two additional tests using delays inside the temperature chamber of 6 h before testing showed a relative increase in the viscosity over the tests with 30 min delays of approximately 1 %. A simple linear model is assumed for the relative change in the viscosity with time at a rate of 1 % for an increased delay of 330 min. This assumption leads to the incorporation of a component of relative standard uncertainty of 0.1 % at the first shear rate of 0.001 s^{-1}, with the relative standard uncertainty increasing linearly with time to 0.3 % at the end of the test. The effects of evaporation on the measurements of the first normal stress difference were not significant compared to the uncertainties for either the 90 min delay tests or the 6 h delay tests. Since the pressure in the fluid increases toward the center of the cone and plate fixture [1-6], the first normal stress difference measurement might be less sensitive to effects at the edge such as solvent evaporation. The measurement of the first normal stress difference is apparently more sensitive to characteristics of the instrument, such as temperature fluctuations or transducer response than to the evaporation of the solvent. For the first normal stress difference, it is assumed that the effects of evaporation are similar to those found for the viscosity, incorporating a component of relative standard uncertainty of 0.1 % at the first shear rate of 0.001 s^{-1}, with the relative standard uncertainty increasing linearly with time to 0.3 % at the end of the test.

3.2.2   Testing at 50 °C

The fixtures were again 50 mm diameter, 0.0197 rad cone and plate, with the cone truncated at 0.048 mm from its vertex. The tests were run using the Steady Rate Sweep template supplied with the ARES software, taking five points per decade. Tests were begun using the more sensitive low range of the torque transducer, sweeping from an initial shear rate of 0.001 s^{-1} up to a shear rate of 25.12 s^{-1}. The instrument was then switched to use the high range of the torque transducer, and a rate sweep from 25.12 s^{-1} to 100 s^{-1} was performed. (The switch was necessary, because the torque at the higher shear rates was sufficient to overload the low range of the transducer.) Measurements at each shear rate were made in both the clockwise and counterclockwise directions, with the shear applied for 20 s and then measurements of the torque and normal force averaged over the next 30 s. Before loading each sample, the cone and plate fixtures were installed and set to a nominal gap of 1 mm. The temperature control chamber was then closed and the fixtures brought to 50 °C for 30 min. The gap was then zeroed using the Autozero capability of the ARES (the fixtures brought together until they touched, to establish the baseline from which to set the gap to the proper dimension). The temperature control chamber was then opened, and the upper fixture was raised approximately 60 mm. Each bottle was turned end over end for at least 30 min before testing to thoroughly mix the material inside; a sample was then spooned onto the lower plate fixture, and the upper cone fixture was lowered to a position of 0.020 mm. The sample was then trimmed flush with the edges of the fixtures, and the cone was repositioned to the specified gap of 0.048 mm. This procedure was used in an effort to compensate for the expansion of the sample when changing from room temperature to the test temperature of 50 °C, in order to achieve the appropriate spherical sample geometry. The temperature control chamber was closed, and the specified Steady Rate Sweep test was started;

the test sequence included a 30 min hold at 50 °C to allow the temperature to reach steady state. Two samples were tested from each of the ten bottles; the twenty samples were tested in a random order.

In reducing the data, the viscosity measurements taken at 25.12 s^{-1} using the low transducer range and the high transducer range were averaged, although typically the difference was much smaller than the uncertainty of the measurements. Averaging measurements made using rotations in both directions eliminates any problems of an offset in the zero position for the viscosity. The normal force measurement, however, is sensitive to the zero position and is strongly affected by temperature fluctuations that cause transient volumetric changes. The zero position of the normal force measurements was adjusted by averaging the first six measurements and subtracting that value from all subsequent measurements. The normal force in that range (0.001 s^{-1} to 0.01 s^{-1}) is well below the sensitivity of the transducer and so can be taken as zero. In changing the transducer from the low torque range to the high torque range, the difference between the two normal force measurements (taken at the same shear rate of 25.12 s^{-1}) was then subtracted from the subsequent measurements taken with the transducer using the high torque range.

The total time for a single test was approximately 80 min, including the 30 min delay to ensure thermal steady state before measurements were begun. A possible source of error in the measurement is evaporation of the solvent during the experiment caused by the forced air temperature control system. To investigate whether evaporation of the solvent caused a significant change in the sample over the duration of the test, three experiments were performed using 80 min delays inside the temperature chamber before measurements were begun, and three experiments were performed using 160 min delays. The results of the 80 min delay tests show a viscosity increase of approximately 1 % over the tests with 30 min delays, and the 160 min delay tests show a viscosity increase of approximately 4 %. A simple linear model is again assumed for the relative change in the viscosity with time at the initial rate of 1 % for an additional delay of 50 min. This assumption leads to the incorporation of a component of relative standard uncertainty of 0.6 % at the first shear rate of 0.001 s^{-1}, with the relative standard uncertainty increasing linearly with time to 1.6 % at the end of the test. Even with the higher rate of evaporation at this temperature, the effects of evaporation on the measurements of the first normal stress difference were not significant compared to the uncertainties for either the 80 min delay tests or the 160 min delay tests. Some additional tests were performed using a series of rate sweep tests on the same specimen, and the effect of evaporation on the viscosity was consistent with the earlier tests with longer delay times. However, the results for the first normal stress difference were still inconclusive. Again, since the pressure in the fluid increases toward the center of the cone and plate fixture [1-6], the first normal stress difference measurement might be less sensitive to effects at the edge such as solvent evaporation. The measurement of the first normal stress difference is apparently more sensitive to characteristics of the instrument, such as temperature fluctuations or transducer response than to the evaporation of the solvent. For the first normal stress difference, we assume that the effects of evaporation are similar to those found for the viscosity, incorporating a component of relative standard uncertainty of 0.6 % at the first shear rate of 0.001 s^{-1}, with the relative standard uncertainty increasing linearly with time to 1.6 % at the end of the test.

3.2.3 Testing at 0 °C

The fixtures were again 50 mm diameter, 0.0197 rad cone and plate, with the cone truncated at 0.048 mm from its vertex. The tests were run using the Steady Rate Sweep template supplied with the ARES software, taking five points per decade. Tests were begun using the more sensitive low range of the torque transducer, sweeping from an initial shear rate of 0.001 s^{-1} to a shear rate of 3.98 s^{-1}. The instrument was then switched to use the high range of the torque transducer, and a rate sweep from 3.98 s^{-1} to 100 s^{-1} was performed. (The switch was necessary, because the torque at the higher shear rates was sufficient to overload the low range of the transducer.) Measurements at each shear rate were made in both the clockwise and counterclockwise directions, with the shear applied for 40 s and then measurements of the torque and normal force averaged over the next 30 s. Before loading each sample, the cone and plate fixtures were installed and set to a nominal gap of 1 mm, and then the temperature control chamber was closed and the fixtures brought to 0 °C for 30 min. Temperatures below room temperature are maintained by evaporated liquid nitrogen into the temperature control chamber. The gap was then zeroed using the Autozero capability of the ARES (the fixtures brought together until they touched, to establish the baseline from which to set the gap to the proper dimension). The temperature control chamber was then opened, and the upper fixture was raised approximately 60 mm. The tool temperature was allowed to reach 15 °C before loading the sample to allow the moisture that condensed on the tool when exposed to the atmosphere to evaporate. Each bottle was turned end over end for at least 30 min before testing to thoroughly mix the material inside; a sample was then spooned onto the lower plate fixture, and the upper cone fixture was lowered to a position of 0.090 mm. The sample was then trimmed flush with the edges of the fixtures. The temperature control chamber was closed, and the sample and fixtures cooled for 4 min. The cone was then repositioned to the specified gap of 0.048 mm. This procedure was used in an effort to compensate for the decrease in sample volume when changing from room temperature to the test temperature of 0 °C, in order to achieve the appropriate spherical sample geometry. The temperature control chamber was closed, and the specified Steady Rate Sweep test was started; the test sequence included a 30 min hold at 0 °C to allow the temperature to reach steady state. Two samples were tested from each of the ten bottles; the twenty samples were tested in a random order.

In reducing the data, the viscosity measurements taken at 3.98 s^{-1} using the low transducer range and the high transducer range were averaged, although typically the difference was much smaller than the uncertainty of the measurements. Averaging measurements made using rotations in both directions eliminates any problems of an offset in the zero position for the viscosity. The normal force measurement, however, is sensitive to the zero position and is strongly affected by temperature fluctuations that cause transient volumetric changes. The zero position of the normal force measurements was adjusted by averaging the first six measurements and subtracting that value from all subsequent measurements. The normal force in that range (0.001 s^{-1} to 0.01 s^{-1}) is well below the sensitivity of the transducer and so can be taken as zero. In changing the transducer from the low torque range to the high torque range, the difference between the two normal force measurements (taken at the same shear rate of 3.98 s^{-1}) was then subtracted from the subsequent measurements taken with the transducer using the high torque range.

Because of the slow rate of change of the viscosity demonstrated by the evaporation tests at 25 °C, no evaporation tests were performed at 0 °C. An estimate of the effect of evaporation on the viscosity and first normal stress difference has been made using the measured rates of change in the viscosity at 25 °C and 50 °C coupled with an Arrhenius temperature dependence. This estimate leads to a component of relative standard uncertainty in both the viscosity and first normal stress difference of 0.01 % at the first shear rate of 0.001 s^{-1}, with the relative standard uncertainty increasing linearly with time to 0.03 % at the end of the test.

3.3 Dynamic Testing

The fixtures used for the dynamic testing were 50 mm diameter parallel plates with the nominal gap being 1 mm. The tests were run using the Dynamic Frequency/Temperature Sweep template supplied with the ARES software. Measurements were taken at 0 °C, 10 °C, 20 °C, 30 °C, 40 °C, and 50 °C; at each temperature the frequency of oscillation increased from 0.0398 rad/s to 100 rad/s, taking five points per decade. Tests were performed at a strain magnitude of 20 %, and all data was taken using the more sensitive low range of the torque transducer. Although the ARES does have some capability to adjust the position of the fixtures to compensate for thermal expansion, this feature is apparently not yet implemented for the Dynamic Frequency/Temperature Sweep test. Therefore, in an effort to minimize the effects of changes in the gap and changes in the sample volume because of thermal expansion of the tools and the sample itself, the gap was zeroed at 25 °C (the midpoint of the temperature range). The change in the gap caused by thermal expansion of the fixtures was measured to be 1.75 μm/°C (with a standard uncertainty of 0.1 μm/°C). Before loading each sample, the parallel plate fixtures were installed and set to a nominal gap of 1 mm, and then the temperature control chamber was closed and the fixtures brought to 25 °C for 30 min. The gap was then zeroed using the Autozero capability of the ARES (the fixtures brought together until they touched, to establish the baseline from which to calculate the gap). The temperature control chamber was then opened, and the upper fixture was raised approximately 60 mm. Each bottle was turned end over end for at least 30 min before testing to thoroughly mix the material inside; a sample was then spooned onto the lower plate fixture, and the upper plate fixture was lowered. The sample was centered in the fixtures by turning off the motor and rotating the lower plate while lowering the upper plate. The upper plate was lowered until the edges of the sample were flush with the edges of the fixtures at room temperature. The gap was then increased 0.01 mm in an effort to compensate for expansion of the tools and the sample for a temperature change from room temperature to 25 °C. The temperature control chamber was closed with the temperature setpoint at 0 °C. The specified Dynamic Frequency/Temperature Sweep test was started; the test sequence included a 30 min hold at 0 °C to allow the temperature to reach steady state. In an effort to reduce the test time somewhat, the subsequent soak time following each 10 °C increment was shortened to 10 min. One sample was tested from each of the ten bottles; the ten samples were tested in a random order. The minimum gap used was 1.115 mm at 25 °C, and the maximum gap used was 1.435 mm at 25 °C.

In reducing the data for G' and G", a correction to the gap was made to account for thermal expansion of the fixtures. Since the fixtures expand with increasing temperature, the gap decreases with increasing temperature. The actual gap is therefore $h_0 - (1.75 \text{ μm/°C})(T - 25 \text{ °C})$, where h_0 is the nominal gap, and T is the temperature at which the measurement is made. The calculated shear moduli are linearly functions of the gap. Therefore, the corrections to the shear moduli at each temperature are given by

$$G'_{corrected} = \left[\frac{h_0 - (1.75\ \mu m/°C)(T - 25°C)}{h_0}\right] G'_{calculated}$$

$$G''_{corrected} = \left[\frac{h_0 - (1.75\ \mu m/°C)(T - 25°C)}{h_0}\right] G''_{calculated}$$

(1)

4. Analysis of Sources of Uncertainty

Uncertainties from each of the sources that are identified are combined through the mathematical formula for the propagation of uncertainties [18]. For a quantity y that is a function of a number of independent quantities x_i, with $y = f(x_i)$, the combined standard uncertainty in y (symbol $u_c(y)$) is calculated from the standard uncertainty in each x_i (symbol $u(x_i)$) as

$$u_c^2(y) = \sum_i \left(\frac{\partial f}{\partial x_i}\right)^2 u^2(x_i) \qquad (2)$$

For input quantities that are not independent, equation (2) would contain terms involving the covariances, but we will assume that the input variables are independent of one another. An alternative way to express equation (2) is that the combined standard uncertainty in y is the summation in quadrature of the components of the standard uncertainty in y arising from each source x_i,

$$u_c^2(y) = \sum_i u^2(y, x_i) \qquad (3)$$

In equation (3), $u(y, x_i)$ is the component of standard uncertainty in y arising from source x_i, which includes Type A uncertainties determined through statistical analysis of the data and Type B uncertainties determined from the formula for the propagation of uncertainty and/or determined by any other means [18].

4.1 Steady Shear Testing in the Cone and Plate Fixtures

There are a number of possible sources of uncertainty in the measurements of the viscosity and first normal stress difference. The Type A uncertainties associated with variability in the material and random influences on the test conditions were assessed through multiple measurements.

Those sources of uncertainty considered to be Type B are listed below.
1. Temperature
2. Transducer
3. Rotation rate
4. Geometry
    a. Gap
    b. Cone angle
    c. Cone/Plate diameter
    d. Cone truncation
    e. Tilt
    f. Concentricity
5. Solvent evaporation
6. Shear heating
7. Inertia-driven secondary flows
8. Edge effects
9. Surface tension

4.1.1 Uncertainty in the Independent Variables: Temperature and Shear Rate

The viscosity and first normal stress difference are tabulated as functions of temperature and shear rate. The standard uncertainty in the temperature is estimated to be 0.1 °C. The advantage of using the cone and plate is that the shear rate is approximately constant throughout the sample. Uncertainty in the shear rate arises from uncertainties in the rotation rate and in the geometry. The nominal shear rate $\dot{\gamma}_0$ in the cone and plate is given by

$$\dot{\gamma}_0 = \frac{\omega_0}{\tan \beta_0} \quad (4)$$

where ω_0 is the specified rate of rotation of the plate, and β_0 is the nominal cone angle (0.0197 rad). The standard uncertainty in the rotation rate $u(\omega)$ is estimated to be $0.005\omega_0$, and the standard uncertainty in the cone angle $u(\beta)$ is estimated to be 10^{-4} rad, based on information from the instrument manufacturer. Measurement of the cone angle at NIST agreed with that given by the manufacturer within that uncertainty. However, it should be noted that Mackay and Dick [19] found much larger uncertainties in a round robin test measuring the geometry of a cone. In equation (4), it is assumed that the cone is perfect and aligned so that the tip would just touch the plate. However, to prevent friction between the fixtures, the tip of the cone is truncated, so there is a small region of decreased shear rate at the center of the fixtures. The manufacturer gives the height of the truncation as 48 μm. There is also uncertainty in the gap between the plate and the position of the tip of the cone. The standard uncertainty in the gap is estimated to be 2.65 μm. Uncertainty in the gap arises from uncertainties in the cone truncation, uncertainty in position at zero gap and at the specified setting, uncertainty in the flatness of the plate and profile of the cone, compliance of the transducer and thermal expansion of the transducer [16, 20-22]. Each of these seven influences was assigned a standard uncertainty of 1 μm. These components of uncertainty were then added in quadrature to calculate the combined standard uncertainty in the gap. Tilting of the axis of the cone with respect to the axis of the plate also introduces uncertainty; the standard uncertainty of the angle of tilt is estimated to be 2×10^{-4} rad. The propagation of uncertainties in ω and β into the uncertainty in $\dot{\gamma}$ can be calculated analytically from equation (4). However, uncertainties associated with the cone truncation, the gap and a tilt lead to effects that vary spatially. The uncertainty in the shear rate is therefore calculated as an average over the area of the plate. Incorporating the effects of the truncation, an offset h_0 from the intended gap and an angle of tilt ϕ, the area-averaged shear rate is given by

$$\bar{\dot{\gamma}} = \frac{1}{\pi R^2} \left[\int_0^{2\pi} \int_0^{r_1} \frac{\omega r^2}{h_0 + r_1 \tan \beta + \phi \sin \theta} dr d\theta + \int_0^{2\pi} \int_{r_1}^{R} \frac{\omega r^2}{h_0 + r \tan \beta + \phi \sin \theta} dr d\theta \right] \quad (5)$$

where $r_1 = 2.45$ mm is the radius of the truncated region, and $R = 25$ mm is the outer radius of the cone or plate. Markovitz et al. [23] examined the effects of geometry on Newtonian viscosity measurements; the terms describing the tilt in equation (5) follow their analysis. The uncertainty associated with each of these sources was calculated separately, using numerical integration [24]. If the angle of tilt is zero, equation (5) reduces to a one-dimensional integral over r; but for nonzero tilt the integral is fully two-dimensional and was calculated by first integrating over r and then over θ. Both the truncation and tilt introduce a bias, which is treated as a standard uncertainty [18]. All of these effects lead to an uncertainty in $\dot{\gamma}$ that is proportional to $\dot{\gamma}$, and the results are given in Table 1 below. It can be seen that the effects of the tilt are negligible.

The combined standard uncertainty is calculated by adding all of the components in quadrature [18].

Table 1. Uncertainty in the Shear Rate	
Source of Uncertainty	Contribution to Standard Uncertainty in $\dot{\gamma}$
Offset in Gap, h_0	$0.010 \times \dot{\gamma}$
Rotation Rate, ω	$0.005 \times \dot{\gamma}$
Cone Angle, β	$0.005 \times \dot{\gamma}$
Cone Truncation	$0.003 \times \dot{\gamma}$
Angle of Tilt, ϕ	$5 \times 10^{-5} \times \dot{\gamma}$
Combined Standard Uncertainty	$0.013 \times \dot{\gamma}$

4.1.2 Uncertainties in the Viscosity and First Normal Stress Difference

The Type A uncertainties were assessed through multiple measurements (twenty samples at each temperature using two samples from each of ten randomly chosen bottles) and statistical analysis. Since the intention is to certify the *mean* values of the viscosity and first normal stress difference as functions of shear rate, the Type A uncertainty is calculated by dividing the standard deviation of the twenty measurements by the square root of the number of measurements, yielding the standard uncertainty of the mean. These components of uncertainty are given in Table 3 for the viscosity and Table 4 for the first normal stress difference.

The effects of the Type B sources of uncertainty listed above are calculated below. The effects of some of these influences will be calculated using functions fit to the experimental data for the viscosity and first normal stress difference. The viscosity $\eta(\dot{\gamma},T)$ has been fit to a Cross model [6, 25], with

$$\eta(\dot{\gamma},T) = \left(\frac{T\rho}{T_R \rho_R}\right) \left(\frac{\eta_R a(T)}{1+(\xi_0 a(T)\dot{\gamma})^{1-n}}\right) \tag{6}$$

where $\dot{\gamma}$ is the shear rate, T is the temperature, ρ is the density at temperature T, η_R is the zero-shear-rate viscosity at the reference temperature $T_R = 25$ °C, ρ_R is the density at the reference temperature T_R, ξ_0 is a parameter that governs the transition from the Newtonian regime at low shear rates to the power law regime at high shear rates, $a(T)$ is the temperature shift factor (discussed below), and n is the power at which the shear stress increases with shear rate. The density is approximated as a linear function of temperature, with $\rho(T) = \rho_R (1 - \alpha(T - T_R))$, where α is the volumetric coefficient of thermal expansion. The volumetric coefficient of thermal expansion is estimated to be $\alpha = 6 \times 10^{-4}$ cm^3/(cm^3 K) [26, 27]. The effect of the change in the density is small compared to the change in the temperature itself.

The temperature dependence is primarily governed by the shift factor $a(T)$, which is equal to the ratio of the zero-shear-rate viscosity at temperature T divided by the zero-shear-rate viscosity at the reference temperature $T_R = 25\ °C$. The shift factor has been fit with a function of the WLF type [6, 25], giving

$$a(T) = \exp\left(\frac{-C_1(T-T_R)}{C_2+T-T_R}\right) \quad (7)$$

The first normal stress difference $N_1(\dot{\gamma},T)$ was also fit to an empirical model similar to the Cross model, but with an additional term in the denominator (using the same temperature shift factor $a(T)$ calculated for the viscosity):

$$N_1(\dot{\gamma},T) = \left(\frac{T\rho}{T_R\rho_R}\right)\left(\frac{\psi_R(a(T)\dot{\gamma})^2}{1+\xi_1 a(T)\dot{\gamma}+(\xi_2 a(T)\dot{\gamma})^p}\right) \quad (8)$$

In equation (8), T is the temperature, ψ_R is the zero-shear-rate first normal stress coefficient at the reference temperature $T_R = 25\ °C$, ρ is the density at temperature T, ρ_R is the density at the reference temperature T_R; ξ_1, ξ_2 and p are parameters fit to the data. Values for the parameters in equations (6) through (8) are given in Table 2.

Table 2. Parameters for $\eta(\dot{\gamma},T)$, $a(T)$ and $N_1(\dot{\gamma},T)$ in the models found in equations (6) through (8).		
Parameter	Value	Standard Uncertainty
η_R	100.2 Pa·s	0.6 Pa·s
ξ_0	0.234 s	0.004 s
n	0.195	0.004
C_1	7.23	0.24
C_2	150 °C	5 °C
ψ_R	129 Pa·s^2	5 Pa·s^2
ξ_1	1.69 s	0.13 s
ξ_2	0.247 s	0.026 s
p	1.67	0.047

The measured viscosity η is calculated from the twisting moment M applied to the transducer as

$$\eta = \frac{3\tan\beta_0}{2\pi R_0^3 \omega_0} M \quad (9)$$

where $R_0 = 25$ mm is the nominal radius of both cone and plate, ω_0 is the nominal rotation rate and $\beta_0 = 0.0197$ rad is the nominal cone angle. These nominal values are taken to be constants used in the calculation, with the effects of the uncertainties in R, ω and β incorporated through their effect on the moment M, which is given by the integral over the area of the plate of the radius times the shear stress (the viscosity multiplying the shear rate):

$$M = \int_0^{2\pi} \int_0^R \eta(\dot{\gamma},T)\dot{\gamma}(r,\theta)r^2 dr d\theta \qquad (10)$$

Markovitz *et al.* [23] assessed the effects of geometric imperfections for Newtonian fluids in closed form, and we have adapted their analysis to the non-Newtonian case. This analysis employs the assumption that the shear rate $\dot{\gamma}$ is given by relative angular velocity of the plate with respect to the cone divided by the gap between them. If the cone and plate are well aligned (concentric and not tilted), the shear rate at a radius r is approximated as

$$\dot{\gamma} = \frac{\omega r}{h_0 + r \tan \beta} \qquad (11)$$

where ω is the rate of rotation of the plate, h_0 is an offset spacing between the cone and the plate, and β is the cone angle. Ideally, the offset h_0 would be zero, in which case the shear rate is independent of the radius r. To achieve the condition $h_0 = 0$ and a constant shear rate throughout most of the sample, the cone is truncated so that the tip of the cone will not touch the plate and cause friction. Approximately 50 μm is truncated from the cone, so the radius of the truncated region is approximately 2.5 mm. The truncation allows the cone to lie above or below its intended position, so that h_0 can be either positive or negative. Within the truncated area, $\beta = 0$, and the shear rate is a linear function of r, as is the case with parallel plate fixtures. To assess the effects of the cone truncation or nonzero h_0, equation (10) is evaluated numerically [24], using the viscosity model in equation (6). In general, geometric imperfections affect the moment in the Newtonian case more strongly than in the non-Newtonian case, because with shear thinning, an error that causes an increase in the shear rate is offset somewhat by a decrease in the viscosity.

The uncertainties in the first normal stress difference are addressed in a similar manner, with N_1 determined from the axial force F applied to the transducer.

$$N_1 = \frac{2}{\pi R_0^2} F \qquad (12)$$

Again, R_0 is taken to be a constant, and the effects of the uncertainties in each parameter on N_1 are calculated through their effect on F, which is calculated through an integral over the area of the plate similar to that in equation (10). Marsh and Pearson [28] have analyzed the axial force generated in the event that a perfect cone is offset from the plate by an amount h_0. The force in that case is given by

$$F = \int_0^{2\pi} \int_0^R \frac{(N_1 - N_2)}{2} r dr d\theta + \int_0^{2\pi} \int_0^R \frac{N_2}{2} \left[\frac{\tan \beta}{h_0 + r \tan \beta} \right] r^2 dr d\theta \qquad (13)$$

where N_2 is the second normal stress difference. The argument for both N_1 and N_2 in equation (13) is the shear rate $\dot{\gamma}$ given by equation (11). When $h_0 = 0$, equation (13) reduces to the result expected for the cone and plate, in which case the terms containing N_2 cancel:

$$F = \int_0^{2\pi} \int_0^R \frac{N_1}{2} r dr d\theta$$
$$= \frac{\pi R^2 N_1}{2} \qquad (14)$$

For parallel plates, where $\beta = 0$, equation (13) reduces to

$$F = \int_0^{2\pi} \int_0^R \frac{(N_1 - N_2)}{2} r\, dr\, d\theta \qquad (15)$$

To assess the effects of the cone truncation on N_1, one could assume that equation (15) holds over the truncated area, while the integral in equation (14) holds over the rest of plate (with the radius of the truncated region as the lower limit in the integration instead of 0). Alternatively, Marsh and Pearson [28] also give an analytical expression for the partial derivative of F with respect to h_0, evaluated at $h_0 = 0$,

$$\frac{\partial F}{\partial h_0} = \frac{-\pi R}{\tan \beta}\left(\dot{\gamma}_0 \frac{\partial N_1}{\partial \dot{\gamma}_0} + N_2\right) \qquad (16)$$

where $\dot{\gamma}_0 = \omega/\tan\beta$ is the nominal shear rate. Tanner [3] gives an equivalent expression in a different form. Note that N_2 is typically opposite in sign and smaller in magnitude than N_1. Since N_1 is a positive, increasing function of $\dot{\gamma}$, a conservative estimate of $\partial F/\partial h_0$ in equation (16) is obtained by taking $N_2 = 0$. Similarly, a conservative estimate of the effect of the cone truncation is obtained by taking $N_2 = -N_1$ in equation (15) applied to the area of the truncation. Alternatively, N_2 can be calculated from measurements made using parallel plates along with the N_1 data measured with the cone and plate fixtures [1-6].

4.1.2.1 Uncertainty η and N_1 Arising from Uncertainty in Temperature

The effects of uncertainty in the temperature can be calculated directly from equations (6) through (8) with the parameters in Table 1. For the viscosity, the component of uncertainty associated with uncertainty in the temperature is given by

$$u(\eta, T) = \frac{\partial \eta}{\partial T} u(T)$$

$$= \eta(\dot{\gamma}, T)\left\{\frac{1}{T} - \frac{\alpha}{1-\alpha(T-T_R)} + \left[\frac{1 + n(\xi_0 a(T)\dot{\gamma})^{1-n}}{1 + (\xi_0 a(T)\dot{\gamma})^{1-n}}\right]\left(\frac{-C_1 C_2}{(C_2 + T - T_R)^2}\right)\right\} u(T) \qquad (17)$$

For the first normal stress difference, the component of uncertainty associated with uncertainty in the temperature is given by

$$u(N_1, T) = \frac{\partial N_1}{\partial T} u(T)$$

$$= N_1(\dot{\gamma}, T)\left\{\frac{1}{T} - \frac{\alpha}{1-\alpha(T-T_R)} + \left[\frac{2 + \xi_1 a(T)\dot{\gamma} + (2-p)(\xi_2 a(T)\dot{\gamma})^p}{1 + \xi_1 a(T)\dot{\gamma} + (\xi_2 a(T)\dot{\gamma})^p}\right]\left(\frac{-C_1 C_2}{(C_2 + T - T_R)^2}\right)\right\} u(T) \qquad (18)$$

The standard uncertainty in the temperature $u(T)$ is estimated to be 0.1 °C. The components of uncertainty in viscosity and first normal stress difference associated with temperature are given in Tables 3 and 4. Note that these uncertainties represent the effects of temperature on the material properties. Temperature fluctuations also affect the measurement of the normal force directly, because the subsequent thermal expansion and contraction of the sample and the fixtures introduce an axial force on the transducer. These temperature fluctuations are accounted for in the Type A uncertainty calculated from the scatter in the data over repeated tests. The temperature fluctuations in the rheometer are much smaller when heating above the ambient temperature than when cooling below the ambient temperature, because the power to the electric heating system can be varied, whereas the cooling is achieved by the flow of cold air that is either on or off. There is a larger scatter in the first normal stress difference at 0 °C compared to the scatter at 25 °C or 50 °C, which is particularly noticeable at low levels of N_1. The scatter in the viscosity data does not show the same dependence on temperature because the thermal fluctuations do not directly affect the measurement of the moment.

4.1.2.2 Uncertainty in η and N_1 Arising from Uncertainty in Transducer Readings

The effects of uncertainties in the transducer readings can be calculated directly from equations (9) and (12). For the viscosity,

$$u(\eta, M) = \frac{3\tan\beta_0}{2\pi R_0^3 \omega_0} u(M) \qquad (19)$$

while for the first normal stress difference,

$$u(N_1, F) = \frac{2}{\pi R_0^2} u(F) \qquad (20)$$

The standard uncertainty in the moment is estimated to be $u(M) = 10^{-7}$ N·m $+ 0.002M$, while the standard uncertainty in the axial force measurement is estimated to be $u(F) = 8\times10^{-4}$ N $+ 0.002F$. These estimates are based on information from the instrument manufacturer and the variability observed in the data. The components of uncertainty in viscosity and first normal stress difference associated with the transducer are given in Tables 3 and 4.

4.1.2.3 Uncertainty in η and N_1 Arising from Uncertainty in Rotation Rate

Uncertainty in the rotation rate affects both the shear rate $\dot\gamma$ and the viscosity in the calculation of the moment M in equation (10). This effect can be calculated using the chain rule to take the derivative of M with respect to $\dot\gamma$ in equation (10), and then take the derivative of $\dot\gamma$ in equation (11) with respect to ω, evaluated at $h_0 = 0$. Using the model in equation (6) for the viscosity, the component of uncertainty associated with uncertainty in the rotation rate is given by

$$u(\eta, \omega) = \frac{\eta(\dot\gamma, T)}{\omega_0}\left[\frac{1 + n(\xi_0 a(T)\dot\gamma)^{1-n}}{1 + (\xi_0 a(T)\dot\gamma)^{1-n}}\right] u(\omega) \qquad (21)$$

The component of uncertainty in the first normal stress difference associated with uncertainty in the rotation rate can be calculated directyl from the model in equation (8), with

$$u(N_1, \omega) = \frac{N_1(\dot\gamma, T)}{\omega_0}\left[\frac{2 + \xi_1 a(T)\dot\gamma + (2-p)(\xi_2 a(T)\dot\gamma)^p}{1 + \xi_1 a(T)\dot\gamma + (\xi_2 a(T)\dot\gamma)^p}\right] u(\omega) \qquad (22)$$

The standard uncertainty in the rotation rate $u(\omega)$ is estimated to be $0.005\omega_0$, based on information from the instrument manufacturer. The components of uncertainty in viscosity and first normal stress difference associated with rotation rate are given in Tables 3 and 4.

4.1.2.4 Uncertainty in η and N_1 Arising from Uncertainties Associated with Geometry

4.1.2.4.1 Gap

The effects of uncertainty in the gap were calculated using numerical solution of equations (10) and (13) for the truncated cone geometry with $h_0 = 1$ μm and $h_0 = -1$ μm in equation (11) outside the area of the truncation. The partial derivatives of η and N_1 with respect to h_0 were then evaluated at $h_0 = 0$ for use in equation (2) with the standard uncertainty in the gap estimated to be 2.65 μm. Uncertainty in the gap arises from uncertainty in the cone truncation, uncertainty in the position at zero gap and at the specified setting, uncertainty in the flatness of the plate and profile of the cone, compliance of the transducer and thermal expansion of the transducer [16, 20-22]. Each of these seven influences was assigned a standard uncertainty of 1 μm. These components of uncertainty were then added in quadrature to calculate the combined standard uncertainty in the gap of 2.6 μm. Equation (16) has been used to estimate the uncertainty in N_1, with $N_2 = 0$. The components of uncertainty in viscosity and first normal stress difference associated with the gap are given in Tables 3 and 4.

4.1.2.4.2 Cone Angle

Effects of uncertainty in the cone angle are evaluated similarly to the effects of uncertainty in the rotation rate. Using the model in equations (6) for the viscosity, the component of uncertainty associated with uncertainty in the cone angle is given by

$$u(\eta, \beta) = \frac{-\eta(\dot{\gamma}, T)}{\cos^2 \beta_0 \tan \beta_0} \left[\frac{1 + n(\xi_0 a(T)\dot{\gamma})^{1-n}}{1 + (\xi_0 a(T)\dot{\gamma})^{1-n}} \right] u(\beta) \quad (23)$$

Using the model in equation (8) for the first normal stress difference, the component of uncertainty associated with uncertainty in the cone angle is given by

$$u(N_1, \beta) = \frac{-N_1(\dot{\gamma}, T)}{\cos^2 \beta_0 \tan \beta_0} \left[\frac{2 + \xi_1 a(T)\dot{\gamma} + (2-p)(\xi_2 a(T)\dot{\gamma})^p}{1 + \xi_1 a(T)\dot{\gamma} + (\xi_2 a(T)\dot{\gamma})^p} \right] u(\beta) \quad (24)$$

Since $\beta_0 = 0.0197$ rad, $\cos^2 \beta_0 \approx 1$ and $\tan \beta_0 \approx \beta_0$. The standard uncertainty in the cone angle $u(\beta)$ is estimated to be 10^{-4} rad, based on information from the instrument manufacturer. Measurement of the cone angle at NIST agreed with that given by the manufacturer within that uncertainty. However, it should be noted that Mackay and Dick [19] found much larger uncertainties in a round robin test measuring the geometry of a cone. The components of uncertainty in viscosity and first normal stress difference associated with cone angle are given in Tables 3 and 4.

4.1.2.4.3 Cone/Plate Diameter

For the viscosity, the effects of uncertainties in the diameter of the cone and/or plate are given by

$$u(\eta, R) = \frac{3\eta(\dot{\gamma}, T)}{R_0} u(R) \quad (25)$$

where R_0 is the specified cone/plate diameter, which is 25 mm. For the first normal stress difference, the effects of uncertainties in the diameter of the cone and/or plate are given by

$$u(N_1, R) = \frac{2N_1(\dot{\gamma}, T)}{R_0} u(R) \quad (26)$$

The standard uncertainty in the radius of the cone or the plate is estimated to be 0.0025 mm. The combined standard uncertainty calculated by adding these two components in quadrature is $u(R) = 0.0035$ mm. The components of uncertainty in viscosity and first normal stress difference associated with cone and/or plate diameter are given in Tables 3 and 4.

4.1.2.4.4 Cone Truncation

The truncation of the cone leads to a region where the shear rate is lower than the nominal shear rate of $\dot{\gamma}_0 = \omega/\tan\beta$, which leads to a decrease in the moment and a decrease in the calculated viscosity. The cone truncation therefore introduces a bias. This bias is expressed in the form of a standard uncertainty to be added in quadrature with the other components. For the viscosity, the relative uncertainty associated with the cone truncation is proportional to the cube of the ratio of the diameter of the truncated region to the diameter of the cone/plate; in this case, that ratio is approximately 1/10, so the relative uncertainty in the viscosity is on the order of 10^{-3}. The component of uncertainty in the viscosity arising from cone truncation has been evaluated numerically using the model in equation (6), and the results are given in Table 3.

For the first normal stress difference, the region of decreased shear rate and the change in the geometry in the truncated region also leads to a bias in the calculated N_1. In this case, the relative standard uncertainty associated with the cone truncation is proportional to the square of the ratio of the diameter of the truncated region to the diameter of the cone/plate, so the relative standard uncertainty in N_1 is on the order of 10^{-2}. The component of uncertainty in the viscosity arising from cone truncation has been evaluated numerically using the model in equation (8), and the results are given in Table 4.

4.1.2.4.5 Tilt

An angle of tilt between the axis of the cone and the axis of the plate also introduces a bias in the viscosity and first normal stress difference. This angle is expressed as a standard uncertainty estimated to be 2×10^{-4} rad, and the effects of such a tilt were determined by numerical calculation of equations (10) and (14) using the models in equations (6) and (8), along with the expression for the shear rate developed by Markovitz et al. [23]. In this case, the shear rate is a function of the angle θ, so the integrals in equations (10) and (14) are fully two-dimensional; the numerical evaluation was broken into repeated one-dimensional integrals, first integrating over r and then over θ. The components of uncertainty in the viscosity and first normal stress difference arising from a tilt are given in Tables 3 and 4.

4.1.2.4.6 Concentricity

An offset between the axis of the cone and the axis of the plate also introduces a bias in the viscosity and first normal stress difference. This offset is expressed as a standard uncertainty estimated to be 25 µm. The effects of such an offset were also estimated by numerical calculation of equations (10) and (14) using the models in equations (6) and (8). Markovitz et al. [23] give an expression for the shear rate as

$$\dot{\gamma}(r,\theta) = \frac{\omega r}{\tan\beta (r^2 + 2br\cos\theta + b^2)^{1/2}} \quad (27)$$

where b is the offset between the axis of the cone and the axis of the plate. Again, the integrals in equations (10) and (14) are fully two-dimensional, and the numerical evaluation was broken into repeated one-dimensional integrals, first integrating over r and then over θ. In this case, the components of uncertainty in the viscosity and first normal stress difference arising from an offset between the axes of the cone and the plate are negligible compared to other components.

4.1.2.5 Uncertainty Associated with Solvent Evaporation

The uncertainties in the viscosity and first normal stress difference caused by solvent evaporation were estimated from experiments described in Section 3. The components of uncertainty in the viscosity and first normal stress difference arising from solvent evaporation are given in Tables 3 and 4.

4.1.2.6 Effects of Shear Heating

Energy dissipation through friction can lead to an increase in the temperature in the sample, which would decrease the viscosity and first normal stress difference. Bird, *et al.* [4] suggest that the maximum increase in the temperature caused by shear heating would be

$$\Delta T_{max} = \frac{\eta \omega^2 R^2}{8k} \tag{28}$$

where k is the thermal conductivity of the fluid. The thermal conductivities of organic liquids are typically in the range from 0.1 W/(m·K) to 0.2 W/(m·K) [27, 29]. An estimate of the thermal conductivity of 2,6,10,14-tetramethylpentadecane can be made from its chemical structure [27], giving $k = 0.121$ W/(m·K) at 0 °C with a linear decrease to $k = 0.115$ W/(m·K) at 50 °C. It is reasonable to assume that the thermal conductivity of the solution is effectively the same as that of the solvent alone, but in any event, the thermal conductivity of polyisobutylene is similar to that of the solvent. Van Krevelen [28] gives the thermal conductivity of polyisobutylene as 0.130 W/(m·K). The temperature rise associated with shear heating is therefore calculated using equation (28) to be less than 0.025 °C over the entire range of conditions studied. Taking this value as a standard uncertainty in the temperature and adding it in quadrature with the standard uncertainty estimated for the temperature control of 0.1 °C, the combined standard uncertainty in the temperature is calculated to be 0.103 °C. The added effects of shear heating are therefore considered negligible.

4.1.2.7 Inertia-Driven Secondary Flows

The analysis of the cone and plate employs the assumption that the fluid travels in circular paths. Inertial effects can introduce instabilities that alter the assumed flow field and affect the resulting measurements. For the viscosity, these effects are proportional to the square of the Reynolds number Re [6], which is given by

$$\mathrm{Re} = \frac{\rho \omega \beta^2 R^2}{\eta_0} \tag{29}$$

where ρ is the fluid density and η_0 is the zero-shear-rate viscosity. The relative change in the moment associated with secondary flows is given [6] as

$$\frac{\Delta M}{M} = 6.1 \times 10^{-4} \, \mathrm{Re}^2 \tag{30}$$

Since the Reynolds number is less than 10^{-4} for all the conditions tested, the effect of secondary flow on the moment (and therefore the viscosity measurement) is negligible.

For the first normal stress difference, inertia causes the fluid to try to flow out of the gap between cone and plate, reducing the axial force. The change in the first normal stress difference is given [6] by

$$\Delta N_1 = -0.15 \rho \omega^2 R^2 \tag{31}$$

This bias in the first normal stress difference is treated as a standard uncertainty to be added in quadrature with the other components of uncertainty. This component of uncertainty is given in Table 4. In the calculation, the density has been conservatively approximated as 1 g/cm^3.

4.1.2.8 Edge Effects

The conditions at the edge of the cone and plate can impact the measurements in several ways [6], but these effects are not easily quantifiable. Perhaps the most significant difficulty is that the fluid can escape from between the cone and plate. One indicator of loss of fluid would be a decrease in the moment with increasing shear rate. This decrease was only observed in three experiments at 0 °C in the step from a shear rate of 63 s^{-1} to a shear rate of 100 s^{-1}. Those three measurements were discarded. The only other evidence of edge effects occurs at the three highest shear rates at all three temperatures, where there is an increase in the relative scatter of the viscosity data. For this reason, data at the three highest shear rates are provided as reference data only, since the sample geometry might not match our assumptions, and the uncertainty in the data cannot be completely quantified.

4.1.2.9 Surface Tension

Surface tension also affects the measurements [6], particularly the first normal stress difference. These effects have not been quantified, so no component of uncertainty has been assigned to the data.

Table 3. Components of Standard Uncertainty in the Viscosity η							
Temperature	Shear Rate	Measured Viscosity	Standard Uncertainty (Type A)	$u(\eta,T)$	$u(\eta,M)$	$u(\eta,\omega)$	$u(\eta,\beta)$
°C	s^{-1}	Pa·s	Pa·s	Pa·s	Pa·s	Pa·s	Pa·s
0.0	0.001000	382.900	1.176	2.604	3.822	1.910	1.939
0.0	0.001585	380.049	0.999	2.596	2.694	1.893	1.922
0.0	0.002512	382.381	0.946	2.584	1.982	1.901	1.931
0.0	0.003981	382.901	0.880	2.567	1.534	1.899	1.929
0.0	0.006310	383.973	0.826	2.543	1.252	1.898	1.927
0.0	0.01000	383.078	0.786	2.508	1.072	1.883	1.912
0.0	0.01585	382.897	0.680	2.460	0.959	1.868	1.897
0.0	0.02512	379.825	0.699	2.392	0.881	1.833	1.861
0.0	0.03981	375.059	0.629	2.299	0.827	1.782	1.809
0.0	0.06310	365.510	0.631	2.174	0.779	1.698	1.724
0.0	0.1000	350.006	0.551	2.011	0.731	1.576	1.601
0.0	0.1585	328.432	0.458	1.807	0.676	1.417	1.439
0.0	0.2512	300.793	0.409	1.565	0.614	1.225	1.244
0.0	0.3981	268.517	0.326	1.297	0.545	1.013	1.029
0.0	0.6310	232.977	0.260	1.021	0.471	0.799	0.811
0.0	1.000	196.534	0.211	0.762	0.396	0.600	0.609
0.0	1.585	161.096	0.155	0.540	0.324	0.430	0.437
0.0	2.512	128.413	0.105	0.366	0.258	0.296	0.301
0.0	3.981	99.449	0.075	0.239	0.200	0.197	0.200
0.0	6.310	75.066	0.051	0.153	0.151	0.128	0.130
0.0	10.00	55.594	0.037	0.097	0.111	0.083	0.084
0.0	15.85	40.259	0.030	0.062	0.081	0.054	0.054
0.0	25.12	28.670	0.036	0.040	0.057	0.035	0.035
0.0	39.81	20.230	0.055	0.026	0.041	0.023	0.023
0.0	63.10	13.876	0.062	0.017	0.028	0.015	0.015
0.0	100.0	9.085	0.045	0.011	0.018	9.0×10^{-3}	9.0×10^{-3}

Table 3. (continued) Components of Standard Uncertainty in the Viscosity η							
Temperature	Shear Rate	Measured Viscosity	$u(\eta, h_0)$	$u(\eta, R)$	Cone Truncation	Tilt	Evaporation
°C	s^{-1}	Pa·s	Pa·s	Pa·s	Pa·s	Pa·s	Pa·s
0.0	0.001000	382.900	3.104	0.161	0.091	0.020	0.042
0.0	0.001585	380.049	3.097	0.160	0.091	0.020	0.045
0.0	0.002512	382.381	3.085	0.161	0.091	0.020	0.048
0.0	0.003981	382.901	3.067	0.161	0.090	0.019	0.052
0.0	0.006310	383.973	3.042	0.161	0.090	0.019	0.055
0.0	0.01000	383.078	3.006	0.161	0.089	0.019	0.058
0.0	0.01585	382.897	2.956	0.161	0.088	0.019	0.060
0.0	0.02512	379.825	2.885	0.160	0.086	0.018	0.063
0.0	0.03981	375.059	2.785	0.158	0.083	0.017	0.064
0.0	0.06310	365.510	2.649	0.154	0.080	0.016	0.065
0.0	0.1000	350.006	2.468	0.147	0.076	0.015	0.066
0.0	0.1585	328.432	2.241	0.138	0.070	0.013	0.065
0.0	0.2512	300.793	1.962	0.126	0.062	0.011	0.062
0.0	0.3981	268.517	1.648	0.113	0.053	8.0×10^{-3}	0.058
0.0	0.6310	232.977	1.314	0.098	0.043	6.0×10^{-3}	0.053
0.0	1.000	196.534	0.995	0.083	0.034	4.0×10^{-3}	0.047
0.0	1.585	161.096	0.714	0.068	0.025	2.0×10^{-3}	0.039
0.0	2.512	128.413	0.489	0.054	0.017	1.0×10^{-3}	0.032
0.0	3.981	99.449	0.323	0.042	0.012	1.0×10^{-3}	0.026
0.0	6.310	75.066	0.208	0.032	8.0×10^{-3}	7.0×10^{-4}	0.020
0.0	10.00	55.594	0.133	0.023	5.1×10^{-3}	4.0×10^{-4}	0.015
0.0	15.85	40.259	0.085	0.017	3.3×10^{-3}	3.0×10^{-4}	0.011
0.0	25.12	28.670	0.055	0.012	2.2×10^{-3}	2.0×10^{-4}	8.1×10^{-3}
0.0	39.81	20.230	0.036	8.0×10^{-3}	1.3×10^{-3}	1.0×10^{-4}	5.9×10^{-3}
0.0	63.10	13.876	0.023	6.0×10^{-3}	9.0×10^{-4}	1.0×10^{-4}	4.3×10^{-3}
0.0	100.0	9.085	0.016	4.0×10^{-3}	5.8×10^{-4}	5.0×10^{-5}	3.1×10^{-3}

Table 3. (continued) Components of Standard Uncertainty in the Viscosity η							
Temperature	Shear Rate	Measured Viscosity	Standard Uncertainty (Type A)	$u(\eta,T)$	$u(\eta,M)$	$u(\eta,\omega)$	$u(\eta,\beta)$
°C	s^{-1}	Pa·s	Pa·s	Pa·s	Pa·s	Pa·s	Pa·s
25.0	0.001000	97.929	0.869	0.454	3.252	0.489	0.497
25.0	0.001585	98.050	0.674	0.454	2.130	0.490	0.497
25.0	0.002512	98.250	0.618	0.453	1.414	0.490	0.498
25.0	0.003981	97.927	0.335	0.452	0.964	0.488	0.496
25.0	0.006310	98.352	0.249	0.451	0.681	0.490	0.497
25.0	0.01000	98.057	0.242	0.449	0.502	0.488	0.495
25.0	0.01585	98.717	0.188	0.446	0.390	0.490	0.497
25.0	0.02512	98.778	0.179	0.442	0.319	0.488	0.496
25.0	0.03981	98.613	0.180	0.436	0.274	0.485	0.492
25.0	0.06310	98.373	0.177	0.428	0.245	0.480	0.487
25.0	0.1000	97.522	0.177	0.417	0.226	0.470	0.477
25.0	0.1585	96.130	0.168	0.401	0.212	0.456	0.463
25.0	0.2512	93.706	0.158	0.380	0.200	0.435	0.441
25.0	0.3981	89.999	0.149	0.353	0.188	0.404	0.411
25.0	0.6310	84.592	0.135	0.318	0.174	0.364	0.369
25.0	1.000	77.608	0.116	0.277	0.158	0.315	0.320
25.0	1.585	69.160	0.102	0.231	0.140	0.260	0.264
25.0	2.512	59.985	0.079	0.183	0.121	0.204	0.207
25.0	3.981	50.557	0.062	0.137	0.102	0.153	0.155
25.0	6.310	41.440	0.053	0.097	0.083	0.110	0.111
25.0	10.00	33.036	0.045	0.066	0.066	0.075	0.077
25.0	15.85	25.605	0.041	0.043	0.051	0.050	0.051
25.0	25.12	19.361	0.040	0.027	0.039	0.033	0.033
25.0	39.81	14.262	0.043	0.017	0.029	0.021	0.022
25.0	63.10	10.220	0.058	0.011	0.020	0.014	0.014
25.0	100.0	7.223	0.056	7.0×10^{-3}	0.014	9.0×10^{-3}	9.0×10^{-3}

Table 3. (continued) Components of Standard Uncertainty in the Viscosity η							
Temperature	Shear Rate	Measured Viscosity	$u(\eta, h_0)$	$u(\eta, R)$	Cone Truncation	Tilt	Evaporation
°C	s^{-1}	Pa·s	Pa·s	Pa·s	Pa·s	Pa·s	Pa·s
25.0	0.001000	97.929	0.802	0.041	0.024	5.1×10^{-3}	0.098
25.0	0.001585	98.050	0.801	0.041	0.024	5.1×10^{-3}	0.105
25.0	0.002512	98.250	0.800	0.041	0.024	5.1×10^{-3}	0.112
25.0	0.003981	97.927	0.799	0.041	0.024	5.1×10^{-3}	0.119
25.0	0.006310	98.352	0.797	0.041	0.023	5.0×10^{-3}	0.125
25.0	0.01000	98.057	0.794	0.041	0.023	5.1×10^{-3}	0.132
25.0	0.01585	98.717	0.789	0.041	0.023	5.0×10^{-3}	0.139
25.0	0.02512	98.778	0.782	0.041	0.023	5.0×10^{-3}	0.145
25.0	0.03981	98.613	0.773	0.041	0.023	4.9×10^{-3}	0.151
25.0	0.06310	98.373	0.760	0.041	0.022	4.8×10^{-3}	0.156
25.0	0.1000	97.522	0.741	0.041	0.022	4.6×10^{-3}	0.160
25.0	0.1585	96.130	0.715	0.040	0.021	4.3×10^{-3}	0.164
25.0	0.2512	93.706	0.679	0.039	0.021	4.1×10^{-3}	0.165
25.0	0.3981	89.999	0.633	0.038	0.019	3.7×10^{-3}	0.165
25.0	0.6310	84.592	0.573	0.036	0.018	3.2×10^{-3}	0.162
25.0	1.000	77.608	0.501	0.033	0.016	2.7×10^{-3}	0.155
25.0	1.585	69.160	0.419	0.029	0.014	2.1×10^{-3}	0.145
25.0	2.512	59.985	0.334	0.025	0.011	1.6×10^{-3}	0.131
25.0	3.981	50.557	0.252	0.021	8.6×10^{-3}	1.0×10^{-3}	0.115
25.0	6.310	41.440	0.180	0.017	6.4×10^{-3}	7.0×10^{-4}	0.098
25.0	10.00	33.036	0.123	0.014	4.5×10^{-3}	4.0×10^{-4}	0.080
25.0	15.85	25.605	0.081	0.011	3.1×10^{-3}	3.0×10^{-4}	0.063
25.0	25.12	19.361	0.052	8.0×10^{-3}	2.0×10^{-3}	1.0×10^{-4}	0.049
25.0	39.81	14.262	0.033	6.0×10^{-3}	1.2×10^{-3}	1.0×10^{-4}	0.037
25.0	63.10	10.220	0.021	4.0×10^{-3}	8.0×10^{-4}	1.0×10^{-4}	0.027
25.0	100.0	7.223	0.014	3.0×10^{-3}	5.3×10^{-4}	5.0×10^{-5}	0.020

Table 3. (continued) Components of Standard Uncertainty in the Viscosity η							
Temperature	Shear Rate	Measured Viscosity	Standard Uncertainty (Type A)	$u(\eta,T)$	$u(\eta,M)$	$u(\eta,\omega)$	$u(\eta,\beta)$
°C	s^{-1}	Pa·s	Pa·s	Pa·s	Pa·s	Pa·s	Pa·s
50.0	0.001000	36.744	0.741	0.125	3.129	0.184	0.186
50.0	0.001585	37.196	0.470	0.125	2.008	0.186	0.189
50.0	0.002512	37.623	0.252	0.125	1.293	0.188	0.191
50.0	0.003981	37.303	0.182	0.125	0.842	0.186	0.189
50.0	0.006310	37.723	0.086	0.125	0.560	0.188	0.191
50.0	0.01000	37.727	0.091	0.125	0.381	0.188	0.191
50.0	0.01585	37.503	0.075	0.124	0.268	0.187	0.190
50.0	0.02512	37.548	0.056	0.124	0.197	0.187	0.190
50.0	0.03981	37.830	0.056	0.123	0.152	0.188	0.191
50.0	0.06310	37.800	0.050	0.122	0.124	0.187	0.190
50.0	0.1000	37.761	0.051	0.121	0.106	0.186	0.189
50.0	0.1585	37.658	0.050	0.119	0.095	0.184	0.187
50.0	0.2512	37.394	0.048	0.116	0.087	0.180	0.183
50.0	0.3981	36.930	0.046	0.112	0.082	0.175	0.178
50.0	0.6310	36.097	0.044	0.106	0.077	0.168	0.170
50.0	1.000	34.770	0.040	0.099	0.073	0.157	0.159
50.0	1.585	32.794	0.036	0.090	0.068	0.142	0.144
50.0	2.512	30.200	0.032	0.079	0.062	0.123	0.125
50.0	3.981	27.049	0.024	0.066	0.055	0.102	0.104
50.0	6.310	23.543	0.020	0.053	0.048	0.081	0.082
50.0	10.00	19.946	0.020	0.040	0.040	0.061	0.062
50.0	15.85	16.408	0.016	0.029	0.033	0.044	0.045
50.0	25.12	13.138	0.016	0.020	0.026	0.030	0.031
50.0	39.81	10.242	0.017	0.013	0.021	0.020	0.021
50.0	63.10	7.751	0.018	8.2×10^{-3}	0.016	0.013	0.014
50.0	100.0	5.721	0.025	5.1×10^{-3}	0.011	9.0×10^{-3}	9.0×10^{-3}

Table 3. (continued) Components of Standard Uncertainty in the Viscosity η							
Temperature	Shear Rate	Measured Viscosity	$u(\eta,h_0)$	$u(\eta,R)$	Cone Truncation	Tilt	Evaporation
°C	s^{-1}	Pa·s	Pa·s	Pa·s	Pa·s	Pa·s	Pa·s
50.0	0.001000	36.744	0.308	0.015	9.0×10^{-3}	2.0×10^{-3}	0.243
50.0	0.001585	37.196	0.307	0.016	9.0×10^{-3}	1.9×10^{-3}	0.258
50.0	0.002512	37.623	0.307	0.016	9.0×10^{-3}	2.0×10^{-3}	0.273
50.0	0.003981	37.303	0.307	0.016	9.0×10^{-3}	2.0×10^{-3}	0.288
50.0	0.006310	37.723	0.306	0.016	9.0×10^{-3}	2.0×10^{-3}	0.303
50.0	0.01000	37.727	0.306	0.016	9.0×10^{-3}	2.0×10^{-3}	0.318
50.0	0.01585	37.503	0.305	0.016	9.0×10^{-3}	1.9×10^{-3}	0.332
50.0	0.02512	37.548	0.304	0.016	8.9×10^{-3}	1.9×10^{-3}	0.346
50.0	0.03981	37.830	0.302	0.016	8.9×10^{-3}	1.9×10^{-3}	0.360
50.0	0.06310	37.800	0.300	0.016	8.8×10^{-3}	1.9×10^{-3}	0.373
50.0	0.1000	37.761	0.297	0.016	8.7×10^{-3}	1.9×10^{-3}	0.386
50.0	0.1585	37.658	0.291	0.016	8.7×10^{-3}	1.9×10^{-3}	0.397
50.0	0.2512	37.394	0.285	0.016	8.5×10^{-3}	1.8×10^{-3}	0.406
50.0	0.3981	36.930	0.275	0.016	8.3×10^{-3}	1.7×10^{-3}	0.413
50.0	0.6310	36.097	0.262	0.015	7.9×10^{-3}	1.6×10^{-3}	0.415
50.0	1.000	34.770	0.244	0.015	7.4×10^{-3}	1.4×10^{-3}	0.413
50.0	1.585	32.794	0.222	0.014	6.9×10^{-3}	1.3×10^{-3}	0.405
50.0	2.512	30.200	0.194	0.013	6.1×10^{-3}	1.0×10^{-3}	0.389
50.0	3.981	27.049	0.164	0.011	5.3×10^{-3}	8.0×10^{-4}	0.364
50.0	6.310	23.543	0.131	0.010	4.4×10^{-3}	6.0×10^{-4}	0.332
50.0	10.00	19.946	0.099	8.0×10^{-3}	3.4×10^{-3}	4.0×10^{-4}	0.292
50.0	15.85	16.408	0.071	7.0×10^{-3}	2.5×10^{-3}	3.0×10^{-4}	0.248
50.0	25.12	13.138	0.049	6.0×10^{-3}	1.8×10^{-3}	2.0×10^{-4}	0.204
50.0	39.81	10.242	0.032	4.0×10^{-3}	1.2×10^{-3}	1.0×10^{-4}	0.163
50.0	63.10	7.751	0.021	3.0×10^{-3}	8.0×10^{-4}	7.0×10^{-5}	0.126
50.0	100.0	5.721	0.013	2.0×10^{-3}	5.1×10^{-4}	4.0×10^{-5}	0.095

Table 4. Components of Standard Uncertainty in the First Normal Stress Difference N_1							
Temperature	Shear Rate	Measured N_1	Standard Uncertainty (Type A)	$u(N_1,T)$	$u(N_1,F)$	$u(N_1,\omega)$	$u(N_1,\beta)$
°C	s^{-1}	Pa	Pa	Pa	Pa	Pa	Pa
0.0	0.1585	16.30	5.41	0.24	0.85	0.12	0.12
0.0	0.2512	43.78	4.69	0.42	0.90	0.30	0.30
0.0	0.3981	91.85	5.40	0.68	1.00	0.57	0.58
0.0	0.6310	149.06	6.54	1.06	1.11	0.85	0.86
0.0	1.000	257.45	6.61	1.60	1.33	1.35	1.37
0.0	1.585	372.45	5.65	2.35	1.56	1.81	1.84
0.0	2.512	572.64	4.98	3.37	1.96	2.60	2.64
0.0	3.981	845.17	4.40	4.70	2.51	3.58	3.63
0.0	6.310	1218.81	6.68	6.36	3.25	4.80	4.88
0.0	10.00	1717.43	5.57	8.34	4.25	6.26	6.36
0.0	15.85	2362.56	5.75	10.59	5.54	7.93	8.06
0.0	25.12	3195.73	7.34	13.03	7.21	9.84	9.99
0.0	39.81	4251.21	11.56	15.61	9.32	11.97	12.16
0.0	63.10	5518.63	19.91	18.30	11.85	14.24	14.46
0.0	100.0	7125.98	26.43	21.12	15.07	16.92	17.18

Table 4. (continued) Components of Standard Uncertainty in the First Normal Stress Difference N_1								
Temperature	Shear Rate	Measured N_1	$u(N_1,h_0)$	$u(N_1,R)$	Truncation	Tilt	Evaporation	Inertia
°C	s^{-1}	Pa	Pa	Pa	Pa	Pa	Pa	Pa
0.0	0.1585	16.30	0.26	5×10^{-3}	0.03	2×10^{-3}	5×10^{-3}	9×10^{-7}
0.0	0.2512	43.78	0.64	0.01	0.07	3×10^{-3}	0.01	2×10^{-6}
0.0	0.3981	91.85	1.22	0.03	0.16	5×10^{-3}	0.02	6×10^{-6}
0.0	0.6310	149.06	1.82	0.04	0.33	8×10^{-3}	0.03	1×10^{-5}
0.0	1.000	257.45	2.90	0.07	0.64	0.01	0.05	4×10^{-5}
0.0	1.585	372.45	3.90	0.10	1.14	0.02	0.09	9×10^{-5}
0.0	2.512	572.64	5.59	0.16	1.92	0.02	0.14	2×10^{-4}
0.0	3.981	845.17	7.70	0.24	3.12	0.03	0.22	6×10^{-4}
0.0	6.310	1218.81	10.33	0.34	4.89	0.04	0.33	1×10^{-3}
0.0	10.00	1717.43	13.48	0.48	7.44	0.05	0.49	4×10^{-3}
0.0	15.85	2362.56	17.07	0.66	10.99	0.06	0.69	9×10^{-3}
0.0	25.12	3195.73	21.17	0.89	15.78	0.07	0.96	0.02
0.0	39.81	4251.21	25.76	1.19	22.01	0.08	1.30	0.06
0.0	63.10	5518.63	30.64	1.55	29.85	0.11	1.72	0.14
0.0	100.0	7125.98	36.42	2.00	39.45	0.12	2.22	0.36

Table 4. (continued) Components of Standard Uncertainty in the First Normal Stress Difference N_1

Temperature	Shear Rate	Measured N_1	Standard Uncertainty (Type A)	$u(N_1,T)$	$u(N_1,F)$	$u(N_1,\omega)$	$u(N_1,\beta)$
°C	s^{-1}	Pa	Pa	Pa	Pa	Pa	Pa
25.0	0.1585	2.35	0.48	0.02	0.82	0.02	0.02
25.0	0.2512	5.51	0.32	0.04	0.83	0.05	0.05
25.0	0.3981	12.92	0.34	0.09	0.84	0.10	0.10
25.0	0.6310	26.54	0.42	0.16	0.87	0.20	0.20
25.0	1.000	50.05	0.39	0.29	0.91	0.34	0.34
25.0	1.585	87.62	0.45	0.47	0.99	0.54	0.55
25.0	2.512	148.18	0.59	0.74	1.11	0.84	0.85
25.0	3.981	236.76	1.87	1.12	1.29	1.23	1.25
25.0	6.310	377.26	3.92	1.64	1.57	1.83	1.86
25.0	10.00	585.07	4.70	2.36	1.99	2.64	2.68
25.0	15.85	880.00	4.88	3.30	2.57	3.71	3.77
25.0	25.12	1280.30	5.39	4.48	3.38	5.02	5.10
25.0	39.81	1800.04	6.42	5.89	4.41	6.53	6.63
25.0	63.10	2462.22	14.06	7.50	5.74	8.22	8.35
25.0	100.0	3318.59	16.77	9.25	7.45	10.16	10.32

Table 4. (continued) Components of Standard Uncertainty in the First Normal Stress Difference N_1

Temperature	Shear Rate	Measured N_1	$u(N_1,h_0)$	$u(N_1,R)$	Truncation	Tilt	Evaporation	Inertia
°C	s^{-1}	Pa	Pa	Pa	Pa	Pa	Pa	Pa
25.0	0.1585	2.35	0.05	7×10^{-4}	1×10^{-3}	3×10^{-4}	5×10^{-3}	9×10^{-7}
25.0	0.2512	5.51	0.10	2×10^{-3}	3×10^{-3}	6×10^{-4}	0.01	2×10^{-6}
25.0	0.3981	12.92	0.22	4×10^{-3}	0.01	1×10^{-3}	0.02	6×10^{-6}
25.0	0.6310	26.54	0.42	7×10^{-3}	0.03	2×10^{-3}	0.05	1×10^{-5}
25.0	1.000	50.05	0.73	0.01	0.07	4×10^{-3}	0.09	4×10^{-5}
25.0	1.585	87.62	1.16	0.02	0.17	5×10^{-3}	0.18	9×10^{-5}
25.0	2.512	148.18	1.80	0.04	0.35	9×10^{-3}	0.31	2×10^{-4}
25.0	3.981	236.76	2.65	0.07	0.66	0.01	0.53	6×10^{-4}
25.0	6.310	377.26	3.93	0.11	1.17	0.02	0.86	1×10^{-3}
25.0	10.00	585.07	5.69	0.16	1.97	0.02	1.37	4×10^{-3}
25.0	15.85	880.00	7.98	0.25	3.20	0.03	2.11	9×10^{-3}
25.0	25.12	1280.30	10.80	0.36	5.01	0.04	3.18	0.02
25.0	39.81	1800.04	14.06	0.50	7.60	0.06	4.65	0.06
25.0	63.10	2462.22	17.70	0.69	11.20	0.07	6.62	0.14
25.0	100.0	3318.59	21.86	0.93	16.06	0.08	9.18	0.36

Table 4. (continued) Components of Standard Uncertainty in the First Normal Stress Difference N_1							
Temperature	Shear Rate	Measured N_1	Standard Uncertainty (Type A)	$u(N_1,T)$	$u(N_1,F)$	$u(N_1,\omega)$	$u(N_1,\beta)$
°C	s^{-1}	Pa	Pa	Pa	Pa	Pa	Pa
50.0	0.3981	2.38	0.17	0.01	0.82	0.02	0.02
50.0	0.6310	5.08	0.20	0.03	0.83	0.04	0.04
50.0	1.000	10.73	0.26	0.06	0.84	0.09	0.09
50.0	1.585	21.47	0.22	0.11	0.86	0.16	0.16
50.0	2.512	41.05	0.43	0.20	0.90	0.28	0.28
50.0	3.981	68.85	1.15	0.32	0.95	0.43	0.43
50.0	6.310	114.48	2.62	0.51	1.04	0.65	0.66
50.0	10.00	202.50	3.69	0.78	1.22	1.06	1.08
50.0	15.85	344.80	3.97	1.16	1.50	1.68	1.71
50.0	25.12	557.75	4.66	1.67	1.93	2.53	2.57
50.0	39.81	860.20	5.33	2.35	2.54	3.65	3.70
50.0	63.10	1269.10	6.36	3.21	3.35	5.01	5.08
50.0	100.0	1796.79	8.63	4.25	4.41	6.56	6.66

Table 4. (continued) Components of Standard Uncertainty in the First Normal Stress Difference N_1							
Temperature	Shear Rate	Measured N_1	$u(N_1,h_0)$	$u(N_1,R)$	Truncation	Tilt	Evaporation	Inertia
°C	s^{-1}	Pa	Pa	Pa	Pa	Pa	Pa	Pa
50.0	0.3981	2.38	0.05	7×10^{-4}	9×10^{-4}	3×10^{-4}	0.03	6×10^{-6}
50.0	0.6310	5.08	0.09	1×10^{-3}	3×10^{-3}	6×10^{-4}	0.06	1×10^{-5}
50.0	1.000	10.73	0.19	3×10^{-3}	0.01	1×10^{-3}	0.13	4×10^{-5}
50.0	1.585	21.47	0.34	6×10^{-3}	0.03	2×10^{-3}	0.28	9×10^{-5}
50.0	2.512	41.05	0.60	0.01	0.07	3×10^{-3}	0.56	2×10^{-4}
50.0	3.981	68.85	0.92	0.02	0.16	5×10^{-3}	1.05	6×10^{-4}
50.0	6.310	114.48	1.40	0.03	0.33	8×10^{-3}	1.87	1×10^{-3}
50.0	10.00	202.50	2.28	0.06	0.63	0.01	3.20	4×10^{-3}
50.0	15.85	344.80	3.61	0.01	1.12	0.02	5.26	9×10^{-3}
50.0	25.12	557.75	5.45	0.16	1.89	0.02	8.38	0.02
50.0	39.81	860.20	7.85	0.24	3.07	0.03	12.97	0.06
50.0	63.10	1269.10	10.77	0.36	4.82	0.05	19.55	0.14
50.0	100.0	1796.79	14.12	0.50	7.34	0.05	28.70	0.36

4.2 Dynamic Testing in Parallel Plates

For the dynamic tests, the uncertainties in the measurements of G' and G" at each temperature and frequency will be estimated. Master curves and the associated shift factors will be calculated using polynomial functions fit to the data in logarithmic space. The curve fitting procedure also provides estimates of the uncertainties in the parameters describing the master curves and shift factors.

The Type A uncertainties associated with variability in the material and random influences on the test conditions were assessed through multiple measurements.

Those sources of uncertainty considered to be Type B are listed below.
1. Temperature
2. Frequency of oscillation
3. Cross-correlation procedure
    a. Transducer reading
    b. Oscillation magnitude
4. Geometry
    a. Gap
    b. Plate diameter
    c. Tilt
    d. Concentricity
5. Solvent evaporation
6. Inertia
7. Edge effects

4.2.1 Uncertainty in the Independent Variables: Temperature and Frequency of Oscillation

The storage modulus G' and loss modulus G" are tabulated as functions of temperature and frequency of oscillation. The standard uncertainty in the temperature is estimated to be 0.1 °C. The standard uncertainty in the frequency of oscillation Ω is estimated to be $10^{-4} \times \Omega$.

4.2.2 Uncertainties in G' and G"

The Type A uncertainty in each modulus is again calculated from the standard deviation of the repeated experiments divided by the square root of the number of measurements, which gives the uncertainty in the mean of the repeated measurements. These components of uncertainty are listed in Tables 6 and 7.

Type B uncertainties are calculated through the propagation of uncertainties formula in equation (2). These calculations require some analysis of the method by which G' and G" are determined from the geometry, the boundary conditions imposed, and the measured transducer output. The fluid is deformed by oscillating the bottom plate with respect to the upper plate. The strain $\gamma(t)$ in the fluid can be described as

$$\gamma(t) = \frac{r}{h}\phi(t) \tag{32}$$
$$= \frac{r}{h}\phi_0 \sin(\Omega t + \varepsilon)$$

where r is the radial position, h is the gap between plates, and $\phi(t)$ describes the angular position of the lower plate with respect to the upper plate. The function $\phi(t)$ is assumed to be a sine wave with magnitude of oscillation ϕ_0, frequency of oscillation Ω, and a possible phase offset of ε between the master driving signal at $\sin \Omega t$ and the resulting oscillation of the plate. The resulting shear stress $\tau(t)$ in the fluid is assumed as

$$\tau(t) = G'\frac{r}{h}\phi_0 \sin(\Omega t + \varepsilon) + G''\frac{r}{h}\phi_0 \cos(\Omega t + \varepsilon) \tag{33}$$
$$= |G^*|\frac{r}{h}\phi_0 \sin(\Omega t + \delta + \varepsilon)$$

where G' is the storage modulus, G'' is the loss modulus, |G*| is the magnitude of the complex modulus and δ is the phase offset between the strain and the stress. These parameters are related, with

$$\begin{array}{ll} G' = |G^*|\cos\delta & |G^*|^2 = (G')^2 + (G'')^2 \\ G'' = |G^*|\sin\delta & \tan\delta = \dfrac{G''}{G'} \end{array} \tag{34}$$

The moment M measured by the transducer is a function of time, and is calculated by integrating over the area of the plate the product of the shear stress and the radial position.

$$M(t) = \int_0^{2\pi}\int_0^R \left[G'\frac{r^3}{h}\phi_0 \sin(\Omega t + \varepsilon) + G''\frac{r^3}{h}\phi_0 \cos(\Omega t + \varepsilon) \right] dr d\theta$$
$$= \frac{\pi\phi_0 R^4}{2h}\left[G'\sin(\Omega t + \varepsilon) + G''\cos(\Omega t + \varepsilon) \right] \tag{35}$$
$$= \frac{\pi\phi_0 R^4}{2h}|G^*|\sin(\Omega t + \delta + \varepsilon)$$
$$= M_0 \sin(\Omega t + \delta + \varepsilon)$$

A cross-correlation procedure is used to calculate the storage modulus G' and loss modulus G'' from the measured moment $M(t)$ and the measured oscillation $\phi(t)$ [6, 24, 30, 31]. In principle, this procedure correlates the moment $M(t)$ with the strain to calculate the in-phase storage modulus and with a signal $\pi/2$ radians out of phase with the strain to calculate the loss modulus. In practice, the cross-correlation procedure is used to calculate the in-phase and out-of-phase components of both the strain and moment compared to master signals. In this procedure, a measured signal is multiplied together with a master signal (either $\sin \Omega t$ or $\cos \Omega t$) and the result integrated over one or more periods of oscillation to calculate the magnitude and phase angle of the measured signal with respect to the master signal. This procedure also acts a filter to remove unwanted harmonics and noise [6, 24, 30]. The resulting four calculations are

$$\Phi_1 = \frac{\Omega}{k\pi} \int_0^{2k\pi/\Omega} \phi(t)\sin\Omega t\, dt = \phi_0 \cos\varepsilon$$

$$\Phi_2 = \frac{\Omega}{k\pi} \int_0^{2k\pi/\Omega} \phi(t)\cos\Omega t\, dt = \phi_0 \sin\varepsilon$$

$$\mathbf{M}_1 = \frac{\Omega}{k\pi} \int_0^{2k\pi/\Omega} M(t)\sin\Omega t\, dt = \frac{\pi\phi_0 R^4}{2h}\left[G'\cos\varepsilon - G''\sin\varepsilon\right]$$

$$\mathbf{M}_2 = \frac{\Omega}{k\pi} \int_0^{2k\pi/\Omega} M(t)\cos\Omega t\, dt = \frac{\pi\phi_0 R^4}{2h}\left[G'\sin\varepsilon + G''\cos\varepsilon\right]$$

(36)

where k is the number of cycles over which the integration is performed. The storage and loss moduli are then given by

$$G' = \frac{2h}{\pi R^4}\left[\frac{\mathbf{M}_1\Phi_1 + \mathbf{M}_2\Phi_2}{\Phi_1^2 + \Phi_2^2}\right]$$

$$G'' = \frac{2h}{\pi R^4}\left[\frac{-\mathbf{M}_1\Phi_2 + \mathbf{M}_2\Phi_1}{\Phi_1^2 + \Phi_2^2}\right]$$

(37)

The four quantities $\mathbf{M}_1$, $\mathbf{M}_2$, Φ_1 and Φ_2 are the fundamental outputs from the measurements and the cross-correlation procedure.

The measurements of $G'(\Omega, T)$ and $G''(\Omega, T)$ are shifted to create master curves through time-frequency superposition using a shift factor function $a(T)$, with

$$G'(\Omega, T) = \frac{T\rho}{T_R \rho_R} G'(a(T)\Omega, T_R)$$

$$G''(\Omega, T) = \frac{T\rho}{T_R \rho_R} G''(a(T)\Omega, T_R)$$

(38)

where $T_R = 25\ °C$ is the reference temperature, and $a(T)$ has the same WLF functional form used in equation (7), giving

$$a(T) = \exp\left(\frac{-C_1(T-T_R)}{C_2 + T - T_R}\right)$$

(39)

The logarithms of the storage and loss moduli have been fit to polynomial functions of the logarithm of the frequency for calculation of the shift factors [25]. The polynomial functions converge more quickly than an expansion in functions associated with the Rouse modes, and the resulting master curves appear to be satisfactory. The data were fit to functions of the form

$$\ln\left(\frac{G'(\Omega,T)}{1\ \text{Pa}}\right) = \ln\left(\frac{T\rho}{T_R\rho_R}\right) + \sum_{i=0}^{4} p_i \left(\ln\left(\frac{a(T)\Omega}{1\ \text{rad/s}}\right)\right)^i$$

$$\ln\left(\frac{G''(\Omega,T)}{1\ \text{Pa}}\right) = \ln\left(\frac{T\rho}{T_R\rho_R}\right) + \sum_{i=0}^{4} q_i \left(\ln\left(\frac{a(T)\Omega}{1\ \text{rad/s}}\right)\right)^i$$

(40)

Both the storage and loss modulus data were fit simultaneously to the functions in equation (40) to determine the parameters C_1 and C_2 in the temperature shift factor $a(T)$. These functional representations for the storage and loss moduli are also used to calculate the propagation of the uncertainties in the temperature and in the frequency of oscillation into the uncertainties in the moduli. The calculated parameters are given in Table 5.

Table 5. Parameters for $G'(\Omega, T)$, $G''(\Omega, T)$ and $a(T)$ in the models found in equations (39) and (40).		
Parameter	Value	Standard Uncertainty
p_0	3.177	0.005
p_1	1.235	0.003
p_2	-0.134	0.001
p_3	2.36×10^{-3}	2.7×10^{-4}
p_4	5.20×10^{-4}	6.1×10^{-5}
q_0	4.196	0.005
q_1	0.720	0.003
q_2	-0.0719	0.0011
q_3	-3.18×10^{-3}	2.6×10^{-4}
q_4	7.06×10^{-4}	6.0×10^{-5}
C_1	8.85	0.30
C_2	192 °C	6 °C

4.2.2.1 Uncertainties Associated with Temperature

The components of the standard uncertainty in G' and G" associated with the standard uncertainty in the temperature can be calculated from the functions in equation (40) using the formula for propagation of uncertainty from equation (2).

$$u(G',T) = \frac{\partial G'}{\partial T}u(T)$$

$$= G'(\Omega,T)\left\{\frac{1}{T} - \frac{\alpha}{1-\alpha(T-T_R)} + \left(\frac{-C_1 C_2}{(C_2+T-T_R)^2}\right)\left[\sum_{i=1}^{4} ip_i\left(\ln\left(\frac{a(T)\Omega}{1\text{ rad/s}}\right)\right)^{i-1}\right]\right\}u(T)$$

$$u(G'',T) = \frac{\partial G''}{\partial T}u(T)$$

$$= G''(\Omega,T)\left\{\frac{1}{T} - \frac{\alpha}{1-\alpha(T-T_R)} + \left(\frac{-C_1 C_2}{(C_2+T-T_R)^2}\right)\left[\sum_{i=1}^{4} iq_i\left(\ln\left(\frac{a(T)\Omega}{1\text{ rad/s}}\right)\right)^{i-1}\right]\right\}u(T)$$

(41)

The standard uncertainty in the temperature is $u(T) = 0.1$ °C. The components of the standard uncertainty in G' and G" associated with the standard uncertainty in the temperature are listed in Tables 6 and 7.

4.2.2.2 Uncertainties Associated with Frequency of Oscillation

The components of the standard uncertainty in G' and G" associated with the standard uncertainty in the frequency of oscillation Ω are also calculated from the functions above fit to the data.

$$u(G',\Omega) = \frac{\partial G'}{\partial \Omega}u(\Omega)$$

$$= \frac{G'(\Omega,T)}{\Omega}\left[\sum_{i=1}^{4} ip_i\left(\ln\left(\frac{a(T)\Omega}{1\text{ rad/s}}\right)\right)^{i-1}\right]u(\Omega)$$

$$u(G'',\Omega) = \frac{\partial G''}{\partial \Omega}u(\Omega)$$

$$= \frac{G''(\Omega,T)}{\Omega}\left[\sum_{i=1}^{4} iq_i\left(\ln\left(\frac{a(T)\Omega}{1\text{ rad/s}}\right)\right)^{i-1}\right]u(\Omega)$$

(42)

The standard uncertainty in the frequency of oscillation is $u(\Omega) = 10^{-4} \times \Omega$. The components of the standard uncertainty in G' and G" associated with the standard uncertainty in the frequency of oscillation are listed in Tables 6 and 7.

4.2.2.3 Uncertainties Associated with Cross-correlation of the Transducer Signal and Oscillation Magnitude

The storage modulus G' and loss modulus G" are calculated from cross-correlating the moment and the magnitude of oscillation with master signals of $\sin \Omega t$ and $\cos \Omega t$, as described in connection with equations (36) and (37) above. The cross-correlation procedure does filter out noise and higher harmonics [6, 24, 30]. It seems reasonable to assume that the unwanted signals are no more likely to be correlated with $\sin \Omega t$ than with $\cos \Omega t$, in which case the standard uncertainty in $\mathbf{M}_1$ should be equal to the standard uncertainty in $\mathbf{M}_2$: $u(\mathbf{M}_1) = u(\mathbf{M}_2)$. Similarly, the standard uncertainties in Φ_1 and Φ_2 should be equal: $u(\Phi_1) = u(\Phi_2)$. For this special case, where $u(\mathbf{M}_1) = u(\mathbf{M}_2)$ and $u(\Phi_1) = u(\Phi_2)$, the propagation of the uncertainties into G' and G" have simple forms. Define $u(\mathbf{M}) \equiv u(\mathbf{M}_1) = u(\mathbf{M}_2)$ and $u(\Phi) \equiv u(\Phi_1) = u(\Phi_2)$. The component of standard uncertainty in G' arising from the standard uncertainties in $\mathbf{M}_1$ and $\mathbf{M}_2$ is

$$u^2(G',\mathbf{M}) \equiv u^2(G',\mathbf{M}_1,\mathbf{M}_2) = \left(\frac{\partial G'}{\partial \mathbf{M}_1}\right)^2 u^2(\mathbf{M}_1) + \left(\frac{\partial G'}{\partial \mathbf{M}_2}\right)^2 u^2(\mathbf{M}_2)$$
$$= \left(\frac{2h}{\pi R^4 \phi_0}\right)^2 u^2(\mathbf{M}) \quad (43)$$

The component of standard uncertainty in G" arising from the standard uncertainties in $\mathbf{M}_1$ and $\mathbf{M}_2$ is the same as that found above for G' (assuming $u(\mathbf{M}_1) = u(\mathbf{M}_2)$ and $u(\Phi_1) = u(\Phi_2)$):

$$u^2(G'',\mathbf{M}) \equiv u^2(G'',\mathbf{M}_1,\mathbf{M}_2) = \left(\frac{2h}{\pi R^4 \phi_0}\right)^2 u^2(\mathbf{M}) \quad (44)$$

Similarly, the components of standard uncertainty in G' and G" arising from the standard uncertainties in Φ_1 and Φ_2 are

$$u^2(G',\Phi) \equiv u^2(G',\Phi_1,\Phi_2) = \left(\frac{|G^*|}{\phi_0}\right)^2 u^2(\Phi)$$
$$u^2(G'',\Phi) \equiv u^2(G'',\Phi_1,\Phi_2) = \left(\frac{|G^*|}{\phi_0}\right)^2 u^2(\Phi) \quad (45)$$

As mentioned above, the cross-correlation procedure does filter out some extraneous signals. The instrument manufacturer indicates that the constant term in the transducer standard uncertainty is decreased by approximately one order of magnitude, so the standard uncertainty in $\mathbf{M}$ is taken as $u(\mathbf{M}) = 10^{-8}$ N·m + $(0.002)M_0$, where M_0 is the magnitude of the sinusoidally varying moment, as defined by equation (35). The instrument manufacturer also indicates that the standard uncertainty in the magnitude of oscillation is $u(\Phi) = (0.0025)\phi_0$. The components of the standard uncertainty in G' and G" associated with the cross correlation procedure are listed in Tables 6 and 7.

4.2.2.4 Uncertainties Associated with Geometry

4.2.2.4.1 Gap

The components of the standard uncertainties in G' and G" associated with the standard uncertainty in the gap h are given by

$$u(G',h) = \frac{G'}{h_0}u(h)$$
$$u(G'',h) = \frac{G''}{h_0}u(h)$$
(46)

where h_0 is the specified gap. Uncertainty in the gap arises from the uncertainty in the position at zero gap and at the specified setting, uncertainty in the flatness of the upper and lower plates, uncertainty in the thermal expansion of the fixtures, compliance of the transducer and thermal expansion of the transducer [16, 20-22]. A deviation from flatness of the parallel plates could result in a single point of contact at the edge, so the standard uncertainty associated with the position at zero gap was assigned a standard uncertainty of 2.5 µm. The standard uncertainty in the thermal expansion of the fixtures is 0.1 µm/°C, and the gap was zeroed at 25 °C, so the maximum temperature variation from the known gap dimension is 25 °C. Thus, a value of 2.5 µm was assigned as a conservative estimate of the standard uncertainty in the gap resulting from uncertainty in the thermal expansion of the fixtures. Each of the other five influences was assigned a standard uncertainty of 1 µm. These components were added in quadrature to calculate a combined standard uncertainty in the gap of $u(h) = 4.2$ µm. The minimum gap used in the tests was 1.071 mm, so a conservative estimate of $u(h)/h_0$ is $(0.0042/1.071) = 0.004$. The components of the standard uncertainty in G' and G" associated with the standard uncertainty in the gap are listed in Tables 6 and 7.

4.2.2.4.2 Plate Diameter

The components of the standard uncertainties in G' and G" associated with the standard uncertainty in the plate diameter are given by

$$u(G',R) = 4\frac{G'}{R_0}u(R)$$
$$u(G'',R) = 4\frac{G''}{R_0}u(R)$$
(47)

where $R_0 = 25$ mm is the specified plate radius. The standard uncertainty in the radius of the upper or lower plate is estimated to be 0.0025 mm. The combined standard uncertainty calculated by adding these two components in quadrature is $u(R) = 0.0035$ mm. The components of the standard uncertainty in G' and G" associated with the standard uncertainty in the plate diameter are listed in Tables 6 and 7.

4.2.2.4.3 Tilt

An angle of tilt between the axis of the upper plate and the axis of the lower plate introduces a bias in the measurement of G' and G". This angle is expressed as a standard uncertainty estimated to be 2×10^{-4} rad, and the effects of such a tilt were determined by a two dimensional numerical solution of the integral in equation (35). For a gap of 1 mm, the tilt introduces a component of standard uncertainty in G' of (0.005)G' and a similar component of standard uncertainty in G" of (0.005)G". The components of the standard uncertainty in G' and G" associated with a tilt are listed in Tables 6 and 7.

4.2.2.4.4 Concentricity

An offset between the axis of the upper plate and the axis of the lower plate also introduces a bias in the measurement of G' and G". This offset is expressed as a standard uncertainty estimated to be 25 μm. As with the case of measurements in steady shear, the uncertainties in G' and G" associated with this amount of offset are negligible.

4.2.2.5 Solvent Evaporation

Solvent evaporation should affect the measurements of G' and G" to an extent that is similar to the effects of solvent evaporation on the steady shear viscosity measurements. An estimate of the effect of evaporation on G' and G" has been made using the measured rates of change in the viscosity at 25 °C and 50 °C coupled with an Arrhenius temperature dependence. The components of standard uncertainty in G' and G" associated with solvent evaporation are listed in Tables 6 and 7.

4.2.2.6 Inertia

Inertial effects have been neglected in the equations describing the dynamics of the fluid and the instrument. These influences are not expected to be large, particularly in a strain-controlled rheometer, but the uncertainties associated with inertial effects have not been quantified, so no component of uncertainty has been assigned to the data.

4.2.2.7 Edge Effects

In the equations describing the response of the fluid, the geometry is assumed to be a perfect cylinder. The exact shape of the fluid edge will be affected by thermal expansion, surface tension and fluid migration, and the shape of the edge will affect the measurements of G' and G". Again, these effects are not expected to be large, but they have not been quantified, so no component of uncertainty has been assigned to the data.

Table 6. Components of Standard Uncertainty in the Storage Modulus G'							
Temperature	Frequency	Storage Modulus G'	Standard Uncertainty (Type A)	$u(G',T)$	$u(G',\Omega)$	$u(G',\mathbf{M})$	$u(G',\Phi)$
°C	rad/s	Pa	Pa	Pa	Pa	Pa	Pa
0.0	0.02512	0.605	7.7×10^{-3}	6.3×10^{-3}	1.1×10^{-4}	0.018	0.020
0.0	0.03981	1.385	9.4×10^{-3}	0.014	2.3×10^{-4}	0.027	0.031
0.0	0.06310	2.973	8.3×10^{-3}	0.027	4.6×10^{-4}	0.040	0.048
0.0	0.1000	6.067	0.012	0.052	8.8×10^{-4}	0.060	0.072
0.0	0.1585	11.699	0.028	0.092	1.6×10^{-3}	0.088	0.107
0.0	0.2512	21.207	0.048	0.152	2.6×10^{-3}	0.126	0.155
0.0	0.3981	36.214	0.066	0.236	4.1×10^{-3}	0.177	0.219
0.0	0.6310	58.511	0.123	0.343	5.9×10^{-3}	0.244	0.302
0.0	1.000	89.636	0.196	0.470	8.2×10^{-3}	0.327	0.406
0.0	1.585	130.948	0.263	0.606	0.011	0.428	0.533
0.0	2.512	182.959	0.400	0.740	0.013	0.547	0.681
0.0	3.981	246.085	0.565	0.858	0.015	0.684	0.852
0.0	6.310	319.791	0.804	0.953	0.017	0.838	1.044
0.0	10.00	403.264	1.035	1.022	0.019	1.007	1.256
0.0	15.85	495.279	1.284	1.075	0.020	1.190	1.485
0.0	25.12	593.784	1.610	1.129	0.022	1.385	1.728
0.0	39.81	697.210	1.966	1.210	0.023	1.589	1.984
0.0	63.10	804.155	2.290	1.355	0.026	1.802	2.250
0.0	100.0	912.794	2.672	1.615	0.031	2.022	2.525

Table 6. (continued) Components of Standard Uncertainty in the Storage Modulus G'						
Temperature	Frequency	Storage Modulus G'	$u(G',h)$	$u(G',R)$	Tilt	Evaporation
°C	rad/s	Pa	Pa	Pa	Pa	Pa
0.0	0.02512	0.605	2.4×10^{-3}	3.4×10^{-4}	3.0×10^{-3}	2.4×10^{-5}
0.0	0.03981	1.385	5.5×10^{-3}	7.8×10^{-4}	6.9×10^{-3}	9.7×10^{-5}
0.0	0.06310	2.973	0.012	1.7×10^{-3}	0.015	2.6×10^{-4}
0.0	0.1000	6.067	0.024	3.4×10^{-3}	0.030	6.0×10^{-4}
0.0	0.1585	11.699	0.047	6.6×10^{-3}	0.058	1.2×10^{-3}
0.0	0.2512	21.207	0.085	0.012	0.106	2.3×10^{-3}
0.0	0.3981	36.214	0.145	0.020	0.181	4.0×10^{-3}
0.0	0.6310	58.511	0.234	0.033	0.293	6.6×10^{-3}
0.0	1.000	89.636	0.359	0.050	0.448	0.010
0.0	1.585	130.948	0.524	0.073	0.655	0.015
0.0	2.512	182.959	0.732	0.102	0.915	0.021
0.0	3.981	246.085	0.984	0.138	1.230	0.029
0.0	6.310	319.791	1.279	0.179	1.599	0.037
0.0	10.00	403.264	1.613	0.226	2.016	0.047
0.0	15.85	495.279	1.981	0.277	2.476	0.058
0.0	25.12	593.784	2.375	0.333	2.969	0.070
0.0	39.81	697.210	2.789	0.390	3.486	0.082
0.0	63.10	804.155	3.217	0.450	4.021	0.095
0.0	100.0	912.794	3.651	0.511	4.564	0.109

| Table 6. (continued) Components of Standard Uncertainty in the Storage Modulus G' ||||||||
Temperature	Frequency	Storage Modulus G'	Standard Uncertainty (Type A)	$u(G',T)$	$u(G',\Omega)$	$u(G',\mathbf{M})$	$u(G',\Phi)$
°C	rad/s	Pa	Pa	Pa	Pa	Pa	Pa
10.0	0.03981	0.486	6.1×10^{-3}	4.7×10^{-3}	9.0×10^{-5}	0.017	0.018
10.0	0.06310	1.099	9.8×10^{-3}	0.010	2.0×10^{-4}	0.025	0.029
10.0	0.1000	2.438	0.012	0.021	4.0×10^{-4}	0.037	0.044
10.0	0.1585	5.076	0.014	0.041	7.8×10^{-4}	0.056	0.067
10.0	0.2512	10.023	0.026	0.073	1.4×10^{-3}	0.082	0.100
10.0	0.3981	18.531	0.041	0.123	2.4×10^{-3}	0.119	0.146
10.0	0.6310	32.329	0.077	0.194	3.8×10^{-3}	0.169	0.209
10.0	1.000	53.217	0.138	0.287	5.6×10^{-3}	0.234	0.290
10.0	1.585	83.154	0.199	0.399	7.8×10^{-3}	0.317	0.394
10.0	2.512	123.313	0.303	0.523	0.010	0.418	0.520
10.0	3.981	174.767	0.393	0.646	0.013	0.538	0.670
10.0	6.310	238.200	0.537	0.757	0.015	0.678	0.846
10.0	10.00	313.204	0.738	0.848	0.017	0.837	1.044
10.0	15.85	398.816	1.002	0.916	0.019	1.013	1.263
10.0	25.12	493.868	1.272	0.965	0.020	1.204	1.502
10.0	39.81	596.352	1.585	1.010	0.022	1.408	1.757
10.0	63.10	704.450	1.921	1.073	0.024	1.623	2.026
10.0	100.0	816.167	2.268	1.186	0.026	1.847	2.306

Table 6. (continued) Components of Standard Uncertainty in the Storage Modulus G'						
Temperature	Frequency	Storage Modulus G'	$u(G',h)$	$u(G',R)$	Tilt	Evaporation
°C	rad/s	Pa	Pa	Pa	Pa	Pa
10.0	0.03981	0.486	1.9×10^{-3}	2.7×10^{-4}	2.4×10^{-3}	1.1×10^{-4}
10.0	0.06310	1.099	4.4×10^{-3}	6.2×10^{-4}	5.5×10^{-3}	3.1×10^{-4}
10.0	0.1000	2.438	9.8×10^{-3}	1.4×10^{-3}	0.012	7.5×10^{-4}
10.0	0.1585	5.076	0.020	2.8×10^{-3}	0.025	1.7×10^{-3}
10.0	0.2512	10.023	0.040	5.6×10^{-3}	0.050	3.4×10^{-3}
10.0	0.3981	18.531	0.074	0.010	0.093	6.4×10^{-3}
10.0	0.6310	32.329	0.129	0.018	0.162	0.011
10.0	1.000	53.217	0.213	0.030	0.266	0.019
10.0	1.585	83.154	0.333	0.047	0.416	0.029
10.0	2.512	123.313	0.493	0.069	0.617	0.044
10.0	3.981	174.767	0.699	0.098	0.874	0.062
10.0	6.310	238.200	0.953	0.133	1.191	0.085
10.0	10.00	313.204	1.253	0.175	1.566	0.112
10.0	15.85	398.816	1.595	0.223	1.994	0.143
10.0	25.12	493.868	1.975	0.277	2.469	0.178
10.0	39.81	596.352	2.385	0.334	2.982	0.216
10.0	63.10	704.450	2.818	0.394	3.522	0.255
10.0	100.0	816.167	3.265	0.457	4.081	0.296

Table 6. (continued) Components of Standard Uncertainty in the Storage Modulus G'							
Temperature	Frequency	Storage Modulus G'	Standard Uncertainty (Type A)	$u(G',T)$	$u(G',\Omega)$	$u(G',\mathbf{M})$	$u(G',\Phi)$
°C	rad/s	Pa	Pa	Pa	Pa	Pa	Pa
20.0	0.03981	0.182	5.8×10^{-3}	1.7×10^{-3}	3.6×10^{-5}	0.011	0.012
20.0	0.06310	0.430	8.6×10^{-3}	4.0×10^{-3}	8.4×10^{-5}	0.017	0.018
20.0	0.1000	1.023	8.8×10^{-3}	8.7×10^{-3}	1.9×10^{-4}	0.025	0.028
20.0	0.1585	2.275	0.012	0.018	3.8×10^{-4}	0.037	0.044
20.0	0.2512	4.809	0.017	0.035	7.4×10^{-4}	0.056	0.067
20.0	0.3981	9.571	0.031	0.063	1.4×10^{-3}	0.082	0.100
20.0	0.6310	17.918	0.050	0.107	2.3×10^{-3}	0.120	0.147
20.0	1.000	31.620	0.084	0.170	3.7×10^{-3}	0.171	0.211
20.0	1.585	52.610	0.129	0.253	5.5×10^{-3}	0.238	0.295
20.0	2.512	82.735	0.188	0.354	7.8×10^{-3}	0.323	0.401
20.0	3.981	123.780	0.274	0.466	0.010	0.428	0.532
20.0	6.310	176.784	0.384	0.580	0.013	0.553	0.689
20.0	10.00	242.540	0.490	0.683	0.015	0.700	0.873
20.0	15.85	320.640	0.637	0.768	0.018	0.867	1.081
20.0	25.12	410.368	0.834	0.831	0.019	1.052	1.313
20.0	39.81	510.091	1.069	0.876	0.021	1.254	1.565
20.0	63.10	617.780	1.312	0.915	0.022	1.470	1.836
20.0	100.0	731.238	1.575	0.968	0.024	1.698	2.120

Table 6. (continued) Components of Standard Uncertainty in the Storage Modulus G'						
Temperature	Frequency	Storage Modulus G'	$u(G',h)$	$u(G',R)$	Tilt	Evaporation
°C	rad/s	Pa	Pa	Pa	Pa	Pa
20.0	0.03981	0.182	7.3×10^{-4}	1.0×10^{-4}	9.1×10^{-4}	1.1×10^{-4}
20.0	0.06310	0.430	1.7×10^{-3}	2.4×10^{-4}	2.2×10^{-3}	3.2×10^{-4}
20.0	0.1000	1.023	4.1×10^{-3}	5.7×10^{-4}	5.1×10^{-3}	8.2×10^{-4}
20.0	0.1585	2.275	9.1×10^{-3}	1.3×10^{-3}	0.011	1.9×10^{-3}
20.0	0.2512	4.809	0.019	2.7×10^{-3}	0.024	4.2×10^{-3}
20.0	0.3981	9.571	0.038	5.4×10^{-3}	0.048	8.5×10^{-3}
20.0	0.6310	17.918	0.072	0.010	0.090	0.016
20.0	1.000	31.620	0.126	0.018	0.158	0.029
20.0	1.585	52.610	0.210	0.029	0.263	0.048
20.0	2.512	82.735	0.331	0.046	0.414	0.076
20.0	3.981	123.780	0.495	0.069	0.619	0.114
20.0	6.310	176.784	0.707	0.099	0.884	0.163
20.0	10.00	242.540	0.970	0.136	1.213	0.224
20.0	15.85	320.640	1.283	0.180	1.603	0.298
20.0	25.12	410.368	1.641	0.230	2.052	0.382
20.0	39.81	510.091	2.040	0.286	2.550	0.476
20.0	63.10	617.780	2.471	0.346	3.089	0.578
20.0	100.0	731.238	2.925	0.409	3.656	0.685

Table 6. (continued) Components of Standard Uncertainty in the Storage Modulus G'							
Temperature	Frequency	Storage Modulus G'	Standard Uncertainty (Type A)	$u(G',T)$	$u(G',\Omega)$	$u(G',\mathbf{M})$	$u(G',\Phi)$
°C	rad/s	Pa	Pa	Pa	Pa	Pa	Pa
30.0	0.03981	0.077	2.9×10^{-3}	6.3×10^{-4}	1.5×10^{-5}	8.1×10^{-3}	7.6×10^{-3}
30.0	0.06310	0.178	7.1×10^{-3}	1.6×10^{-3}	3.7×10^{-5}	0.012	0.012
30.0	0.1000	0.440	7.5×10^{-3}	3.7×10^{-3}	8.6×10^{-5}	0.017	0.019
30.0	0.1585	1.039	8.1×10^{-3}	8.0×10^{-3}	1.9×10^{-4}	0.025	0.029
30.0	0.2512	2.323	0.012	0.017	3.9×10^{-4}	0.038	0.045
30.0	0.3981	4.899	0.021	0.032	7.6×10^{-4}	0.057	0.069
30.0	0.6310	9.776	0.037	0.058	1.4×10^{-3}	0.085	0.103
30.0	1.000	18.376	0.058	0.099	2.4×10^{-3}	0.124	0.152
30.0	1.585	32.459	0.094	0.157	3.8×10^{-3}	0.176	0.218
30.0	2.512	54.041	0.139	0.234	5.7×10^{-3}	0.246	0.305
30.0	3.981	85.240	0.186	0.327	8.0×10^{-3}	0.334	0.415
30.0	6.310	127.700	0.262	0.430	0.011	0.443	0.552
30.0	10.00	182.644	0.340	0.534	0.013	0.574	0.715
30.0	15.85	250.728	0.433	0.628	0.016	0.726	0.905
30.0	25.12	331.736	0.548	0.706	0.018	0.900	1.122
30.0	39.81	424.451	0.663	0.763	0.020	1.092	1.363
30.0	63.10	527.345	0.798	0.803	0.022	1.302	1.626
30.0	100.0	638.186	0.947	0.838	0.023	1.527	1.906

Table 6. (continued) Components of Standard Uncertainty in the Storage Modulus G'						
Temperature	Frequency	Storage Modulus G'	$u(G',h)$	$u(G',R)$	Tilt	Evaporation
°C	rad/s	Pa	Pa	Pa	Pa	Pa
30.0	0.03981	0.077	3.1×10^{-4}	4.3×10^{-5}	3.8×10^{-4}	1.2×10^{-4}
30.0	0.06310	0.178	7.1×10^{-4}	1.0×10^{-4}	8.9×10^{-4}	3.2×10^{-4}
30.0	0.1000	0.440	1.8×10^{-3}	2.5×10^{-4}	2.2×10^{-3}	8.5×10^{-4}
30.0	0.1585	1.039	4.2×10^{-3}	5.8×10^{-4}	5.2×10^{-3}	2.1×10^{-3}
30.0	0.2512	2.323	9.3×10^{-3}	1.3×10^{-3}	0.012	4.8×10^{-3}
30.0	0.3981	4.899	0.020	2.7×10^{-3}	0.024	0.010
30.0	0.6310	9.776	0.039	5.5×10^{-3}	0.049	0.021
30.0	1.000	18.376	0.074	0.010	0.092	0.040
30.0	1.585	32.459	0.130	0.018	0.162	0.071
30.0	2.512	54.041	0.216	0.030	0.270	0.118
30.0	3.981	85.240	0.341	0.048	0.426	0.186
30.0	6.310	127.700	0.511	0.072	0.639	0.280
30.0	10.00	182.644	0.731	0.102	0.913	0.401
30.0	15.85	250.728	1.003	0.140	1.254	0.553
30.0	25.12	331.736	1.327	0.186	1.659	0.733
30.0	39.81	424.451	1.698	0.238	2.122	0.940
30.0	63.10	527.345	2.109	0.295	2.637	1.171
30.0	100.0	638.186	2.553	0.357	3.191	1.419

| Table 6. (continued) Components of Standard Uncertainty in the Storage Modulus G' ||||||||
Temperature	Frequency	Storage Modulus G'	Standard Uncertainty (Type A)	$u(G',T)$	$u(G',\Omega)$	$u(G',\mathbf{M})$	$u(G',\Phi)$
°C	rad/s	Pa	Pa	Pa	Pa	Pa	Pa
40.0	0.03981	0.030	4.9×10^{-3}	2.4×10^{-4}	6.3×10^{-6}	6.1×10^{-3}	5.1×10^{-3}
40.0	0.06310	0.062	4.7×10^{-3}	6.4×10^{-4}	1.7×10^{-5}	8.5×10^{-3}	8.1×10^{-3}
40.0	0.1000	0.200	6.4×10^{-3}	1.6×10^{-3}	4.1×10^{-5}	0.012	0.013
40.0	0.1585	0.478	3.1×10^{-3}	3.7×10^{-3}	9.6×10^{-5}	0.018	0.020
40.0	0.2512	1.142	8.8×10^{-3}	8.0×10^{-3}	2.1×10^{-4}	0.027	0.031
40.0	0.3981	2.562	0.015	0.016	4.3×10^{-4}	0.041	0.048
40.0	0.6310	5.391	0.020	0.032	8.3×10^{-4}	0.061	0.074
40.0	1.000	10.695	0.035	0.057	1.5×10^{-3}	0.090	0.110
40.0	1.585	19.966	0.054	0.096	2.6×10^{-3}	0.131	0.161
40.0	2.512	35.035	0.084	0.152	4.1×10^{-3}	0.187	0.231
40.0	3.981	58.033	0.121	0.224	6.0×10^{-3}	0.260	0.322
40.0	6.310	91.010	0.165	0.311	8.4×10^{-3}	0.353	0.438
40.0	10.00	135.717	0.218	0.407	0.011	0.467	0.581
40.0	15.85	193.163	0.289	0.503	0.014	0.602	0.750
40.0	25.12	264.093	0.369	0.588	0.017	0.761	0.948
40.0	39.81	347.741	0.469	0.657	0.019	0.940	1.172
40.0	63.10	443.102	0.576	0.707	0.021	1.139	1.421
40.0	100.0	548.169	0.708	0.742	0.022	1.355	1.691

Table 6. (continued) Components of Standard Uncertainty in the Storage Modulus G'						
Temperature	Frequency	Storage Modulus G'	$u(G',h)$	$u(G',R)$	Tilt	Evaporation
°C	rad/s	Pa	Pa	Pa	Pa	Pa
40.0	0.03981	0.030	1.2×10^{-4}	1.7×10^{-5}	1.5×10^{-4}	1.1×10^{-4}
40.0	0.06310	0.062	2.5×10^{-4}	3.4×10^{-5}	3.1×10^{-4}	2.5×10^{-4}
40.0	0.1000	0.200	8.0×10^{-4}	1.1×10^{-4}	1.0×10^{-3}	8.7×10^{-4}
40.0	0.1585	0.478	1.9×10^{-3}	2.7×10^{-4}	2.4×10^{-3}	2.2×10^{-3}
40.0	0.2512	1.142	4.6×10^{-3}	6.4×10^{-4}	5.7×10^{-3}	5.3×10^{-3}
40.0	0.3981	2.562	0.010	1.4×10^{-3}	0.013	0.012
40.0	0.6310	5.391	0.022	3.0×10^{-3}	0.027	0.026
40.0	1.000	10.695	0.043	6.0×10^{-3}	0.053	0.052
40.0	1.585	19.966	0.080	0.011	0.100	0.098
40.0	2.512	35.035	0.140	0.020	0.175	0.172
40.0	3.981	58.033	0.232	0.032	0.290	0.285
40.0	6.310	91.010	0.364	0.051	0.455	0.449
40.0	10.00	135.717	0.543	0.076	0.679	0.671
40.0	15.85	193.163	0.773	0.108	0.966	0.957
40.0	25.12	264.093	1.056	0.148	1.320	1.312
40.0	39.81	347.741	1.391	0.195	1.739	1.732
40.0	63.10	443.102	1.772	0.248	2.216	2.211
40.0	100.0	548.169	2.193	0.307	2.741	2.740

Table 6. (continued) Components of Standard Uncertainty in the Storage Modulus G'							
Temperature	Frequency	Storage Modulus G'	Standard Uncertainty (Type A)	$u(G',T)$	$u(G',\Omega)$	$u(G',\mathbf{M})$	$u(G',\Phi)$
°C	rad/s	Pa	Pa	Pa	Pa	Pa	Pa
50.0	0.03981	0.013	5.6×10^{-3}	9.9×10^{-5}	2.8×10^{-6}	5.0×10^{-3}	3.6×10^{-3}
50.0	0.06310	0.034	5.9×10^{-3}	2.7×10^{-4}	7.7×10^{-6}	6.7×10^{-3}	5.8×10^{-3}
50.0	0.1000	0.088	4.7×10^{-3}	7.0×10^{-4}	2.0×10^{-5}	9.3×10^{-3}	9.1×10^{-3}
50.0	0.1585	0.241	4.1×10^{-3}	1.7×10^{-3}	4.9×10^{-5}	0.013	0.014
50.0	0.2512	0.592	5.2×10^{-3}	3.9×10^{-3}	1.1×10^{-4}	0.020	0.022
50.0	0.3981	1.375	0.011	8.5×10^{-3}	2.5×10^{-4}	0.030	0.035
50.0	0.6310	3.058	0.013	0.017	5.0×10^{-4}	0.045	0.054
50.0	1.000	6.385	0.027	0.033	9.6×10^{-4}	0.068	0.082
50.0	1.585	12.509	0.041	0.059	1.7×10^{-3}	0.100	0.122
50.0	2.512	23.028	0.067	0.097	2.9×10^{-3}	0.145	0.179
50.0	3.981	39.969	0.105	0.152	4.5×10^{-3}	0.205	0.254
50.0	6.310	65.434	0.148	0.222	6.6×10^{-3}	0.284	0.353
50.0	10.00	101.548	0.207	0.305	9.1×10^{-3}	0.384	0.477
50.0	15.85	149.841	0.280	0.394	0.012	0.505	0.629
50.0	25.12	211.567	0.363	0.482	0.015	0.649	0.809
50.0	39.81	286.640	0.458	0.558	0.017	0.816	1.018
50.0	63.10	374.506	0.576	0.618	0.020	1.004	1.253
50.0	100.0	473.567	0.703	0.661	0.022	1.212	1.512

Table 6. (continued) Components of Standard Uncertainty in the Storage Modulus G'						
Temperature	Frequency	Storage Modulus G'	$u(G',h)$	$u(G',R)$	Tilt	Evaporation
°C	rad/s	Pa	Pa	Pa	Pa	Pa
50.0	0.03981	0.013	5.1×10^{-5}	7.1×10^{-6}	6.3×10^{-5}	9.7×10^{-5}
50.0	0.06310	0.034	1.4×10^{-4}	1.9×10^{-5}	1.7×10^{-4}	3.0×10^{-4}
50.0	0.1000	0.088	3.5×10^{-4}	4.9×10^{-5}	4.4×10^{-4}	8.3×10^{-4}
50.0	0.1585	0.241	9.7×10^{-4}	1.4×10^{-4}	1.2×10^{-3}	2.4×10^{-3}
50.0	0.2512	0.592	2.4×10^{-3}	3.3×10^{-4}	3.0×10^{-3}	6.0×10^{-3}
50.0	0.3981	1.375	5.5×10^{-3}	7.7×10^{-4}	6.9×10^{-3}	0.014
50.0	0.6310	3.058	0.012	1.7×10^{-3}	0.015	0.032
50.0	1.000	6.385	0.026	3.6×10^{-3}	0.032	0.067
50.0	1.585	12.509	0.050	7.0×10^{-3}	0.063	0.132
50.0	2.512	23.028	0.092	0.013	0.115	0.243
50.0	3.981	39.969	0.160	0.022	0.200	0.424
50.0	6.310	65.434	0.262	0.037	0.327	0.695
50.0	10.00	101.548	0.406	0.057	0.508	1.081
50.0	15.85	149.841	0.599	0.084	0.749	1.600
50.0	25.12	211.567	0.846	0.118	1.058	2.264
50.0	39.81	286.640	1.147	0.161	1.433	3.075
50.0	63.10	374.506	1.498	0.210	1.873	4.025
50.0	100.0	473.567	1.894	0.265	2.368	5.098

Table 7. Components of Standard Uncertainty in the Loss Modulus G"							
Temperature	Frequency	Loss Modulus G"	Standard Uncertainty (Type A)	$u(G'',T)$	$u(G'',\Omega)$	$u(G'',\mathbf{M})$	$u(G'',\Phi)$
°C	rad/s	Pa	Pa	Pa	Pa	Pa	Pa
0.0	0.02512	7.976	0.019	0.045	7.8×10^{-4}	0.018	0.020
0.0	0.03981	12.346	0.031	0.068	1.2×10^{-3}	0.027	0.031
0.0	0.06310	18.852	0.047	0.099	1.7×10^{-3}	0.040	0.048
0.0	0.1000	28.201	0.067	0.139	2.4×10^{-3}	0.060	0.072
0.0	0.1585	41.113	0.098	0.188	3.3×10^{-3}	0.088	0.107
0.0	0.2512	58.262	0.143	0.245	4.3×10^{-3}	0.126	0.155
0.0	0.3981	79.846	0.202	0.303	5.4×10^{-3}	0.177	0.219
0.0	0.6310	105.855	0.300	0.358	6.4×10^{-3}	0.244	0.302
0.0	1.000	135.566	0.360	0.402	7.3×10^{-3}	0.327	0.406
0.0	1.585	168.058	0.481	0.427	7.9×10^{-3}	0.428	0.533
0.0	2.512	201.883	0.592	0.429	8.1×10^{-3}	0.547	0.681
0.0	3.981	235.898	0.724	0.406	7.9×10^{-3}	0.684	0.852
0.0	6.310	268.794	0.871	0.363	7.4×10^{-3}	0.838	1.044
0.0	10.00	299.722	0.994	0.305	6.6×10^{-3}	1.007	1.256
0.0	15.85	328.134	1.122	0.244	5.8×10^{-3}	1.190	1.485
0.0	25.12	354.144	1.244	0.191	5.0×10^{-3}	1.385	1.728
0.0	39.81	378.885	1.361	0.157	4.6×10^{-3}	1.589	1.984
0.0	63.10	404.134	1.499	0.152	4.6×10^{-3}	1.802	2.250
0.0	100.0	432.131	1.685	0.189	5.3×10^{-3}	2.022	2.525

Table 7. (continued) Components of Standard Uncertainty in the Loss Modulus G"						
Temperature	Frequency	Loss Modulus G"	$u(G",h)$	$u(G",R)$	Tilt	Evaporation
°C	rad/s	Pa	Pa	Pa	Pa	Pa
0.0	0.02512	7.976	0.032	4.5×10^{-3}	0.040	3.2×10^{-4}
0.0	0.03981	12.346	0.049	6.9×10^{-3}	0.062	8.6×10^{-4}
0.0	0.06310	18.852	0.075	0.011	0.094	1.6×10^{-3}
0.0	0.1000	28.201	0.113	0.016	0.141	2.8×10^{-3}
0.0	0.1585	41.113	0.164	0.023	0.206	4.3×10^{-3}
0.0	0.2512	58.262	0.233	0.033	0.291	6.4×10^{-3}
0.0	0.3981	79.846	0.319	0.045	0.399	8.9×10^{-3}
0.0	0.6310	105.855	0.423	0.059	0.529	0.012
0.0	1.000	135.566	0.542	0.076	0.678	0.016
0.0	1.585	168.058	0.672	0.094	0.840	0.019
0.0	2.512	201.883	0.808	0.113	1.009	0.023
0.0	3.981	235.898	0.944	0.132	1.179	0.027
0.0	6.310	268.794	1.075	0.151	1.344	0.031
0.0	10.00	299.722	1.199	0.168	1.499	0.035
0.0	15.85	328.134	1.313	0.184	1.641	0.039
0.0	25.12	354.144	1.417	0.198	1.771	0.042
0.0	39.81	378.885	1.516	0.212	1.894	0.045
0.0	63.10	404.134	1.617	0.226	2.021	0.048
0.0	100.0	432.131	1.729	0.242	2.161	0.051

Table 7. (continued) Components of Standard Uncertainty in the Loss Modulus G"								
Temperature	Frequency	Loss Modulus G"	Standard Uncertainty (Type A)	$u(G'',T)$	$u(G'',\Omega)$	$u(G'',M)$	$u(G'',\Phi)$
°C	rad/s	Pa	Pa	Pa	Pa	Pa	Pa
10.0	0.03981	7.295	0.018	0.037	7.2×10^{-4}	0.017	0.018
10.0	0.06310	11.352	0.031	0.056	1.1×10^{-3}	0.025	0.029
10.0	0.1000	17.419	0.044	0.082	1.6×10^{-3}	0.037	0.044
10.0	0.1585	26.287	0.066	0.117	2.3×10^{-3}	0.056	0.067
10.0	0.2512	38.734	0.098	0.160	3.2×10^{-3}	0.082	0.100
10.0	0.3981	55.454	0.138	0.210	4.2×10^{-3}	0.119	0.146
10.0	0.6310	76.894	0.197	0.264	5.3×10^{-3}	0.169	0.209
10.0	1.000	103.161	0.278	0.315	6.4×10^{-3}	0.234	0.290
10.0	1.585	133.741	0.357	0.357	7.3×10^{-3}	0.317	0.394
10.0	2.512	167.622	0.468	0.384	8.0×10^{-3}	0.418	0.520
10.0	3.981	203.424	0.568	0.391	8.3×10^{-3}	0.538	0.670
10.0	6.310	240.105	0.688	0.375	8.3×10^{-3}	0.678	0.846
10.0	10.00	276.093	0.826	0.338	7.8×10^{-3}	0.837	1.044
10.0	15.85	310.338	0.970	0.287	7.0×10^{-3}	1.013	1.263
10.0	25.12	342.151	1.099	0.231	6.1×10^{-3}	1.204	1.502
10.0	39.81	371.952	1.224	0.178	5.3×10^{-3}	1.408	1.757
10.0	63.10	400.674	1.369	0.141	4.8×10^{-3}	1.623	2.026
10.0	100.0	429.896	1.547	0.129	4.7×10^{-3}	1.847	2.306

Table 7. (continued) Components of Standard Uncertainty in the Loss Modulus G"						
Temperature	Frequency	Loss Modulus G"	$u(G",h)$	$u(G",R)$	Tilt	Evaporation
°C	rad/s	Pa	Pa	Pa	Pa	Pa
10.0	0.03981	7.295	0.029	4.1×10^{-3}	0.036	1.7×10^{-3}
10.0	0.06310	11.352	0.045	6.4×10^{-3}	0.057	3.2×10^{-3}
10.0	0.1000	17.419	0.070	9.8×10^{-3}	0.087	5.3×10^{-3}
10.0	0.1585	26.287	0.105	0.015	0.131	8.6×10^{-3}
10.0	0.2512	38.734	0.155	0.022	0.194	0.013
10.0	0.3981	55.454	0.222	0.031	0.277	0.019
10.0	0.6310	76.894	0.308	0.043	0.384	0.027
10.0	1.000	103.161	0.413	0.058	0.516	0.036
10.0	1.585	133.741	0.535	0.075	0.669	0.047
10.0	2.512	167.622	0.670	0.094	0.838	0.060
10.0	3.981	203.424	0.814	0.114	1.017	0.072
10.0	6.310	240.105	0.960	0.134	1.201	0.086
10.0	10.00	276.093	1.104	0.155	1.380	0.099
10.0	15.85	310.338	1.241	0.174	1.552	0.112
10.0	25.12	342.151	1.369	0.192	1.711	0.123
10.0	39.81	371.952	1.488	0.208	1.860	0.134
10.0	63.10	400.674	1.603	0.224	2.003	0.145
10.0	100.0	429.896	1.720	0.241	2.149	0.156

Table 7. (continued) Components of Standard Uncertainty in the Loss Modulus G"							
Temperature	Frequency	Loss Modulus G"	Standard Uncertainty (Type A)	$u(G",T)$	$u(G",\Omega)$	$u(G",\mathbf{M})$	$u(G",\Phi)$
°C	rad/s	Pa	Pa	Pa	Pa	Pa	Pa
20.0	0.03981	4.605	0.011	0.021	4.6×10^{-4}	0.011	0.012
20.0	0.06310	7.259	0.017	0.032	7.1×10^{-4}	0.017	0.018
20.0	0.1000	11.277	0.029	0.049	1.1×10^{-3}	0.025	0.028
20.0	0.1585	17.376	0.041	0.073	1.6×10^{-3}	0.037	0.044
20.0	0.2512	26.304	0.063	0.103	2.3×10^{-3}	0.056	0.067
20.0	0.3981	38.913	0.086	0.142	3.2×10^{-3}	0.082	0.100
20.0	0.6310	56.012	0.133	0.187	4.2×10^{-3}	0.120	0.147
20.0	1.000	78.152	0.173	0.236	5.3×10^{-3}	0.171	0.211
20.0	1.585	105.458	0.238	0.283	6.4×10^{-3}	0.238	0.295
20.0	2.512	137.487	0.307	0.323	7.4×10^{-3}	0.323	0.401
20.0	3.981	173.320	0.390	0.349	8.2×10^{-3}	0.428	0.532
20.0	6.310	211.561	0.482	0.357	8.6×10^{-3}	0.553	0.689
20.0	10.00	250.993	0.577	0.344	8.5×10^{-3}	0.700	0.873
20.0	15.85	290.060	0.682	0.311	8.1×10^{-3}	0.867	1.081
20.0	25.12	327.457	0.792	0.265	7.3×10^{-3}	1.052	1.313
20.0	39.81	362.955	0.909	0.212	6.4×10^{-3}	1.254	1.565
20.0	63.10	396.760	1.036	0.163	5.6×10^{-3}	1.470	1.836
20.0	100.0	429.671	1.199	0.126	5.0×10^{-3}	1.698	2.120

Table 7. (continued) Components of Standard Uncertainty in the Loss Modulus G"						
Temperature	Frequency	Loss Modulus G"	$u(G'',h)$	$u(G'',R)$	Tilt	Evaporation
°C	rad/s	Pa	Pa	Pa	Pa	Pa
20.0	0.03981	4.605	0.018	2.6×10^{-3}	0.023	2.9×10^{-3}
20.0	0.06310	7.259	0.029	4.1×10^{-3}	0.036	5.3×10^{-3}
20.0	0.1000	11.277	0.045	6.3×10^{-3}	0.056	9.1×10^{-3}
20.0	0.1585	17.376	0.070	9.7×10^{-3}	0.087	0.015
20.0	0.2512	26.304	0.105	0.015	0.132	0.023
20.0	0.3981	38.913	0.156	0.022	0.195	0.035
20.0	0.6310	56.012	0.224	0.031	0.280	0.051
20.0	1.000	78.152	0.313	0.044	0.391	0.071
20.0	1.585	105.458	0.422	0.059	0.527	0.096
20.0	2.512	137.487	0.550	0.077	0.687	0.126
20.0	3.981	173.320	0.693	0.097	0.867	0.160
20.0	6.310	211.561	0.846	0.118	1.058	0.195
20.0	10.00	250.993	1.004	0.141	1.255	0.232
20.0	15.85	290.060	1.160	0.162	1.450	0.269
20.0	25.12	327.457	1.310	0.183	1.637	0.305
20.0	39.81	362.955	1.452	0.203	1.815	0.339
20.0	63.10	396.760	1.587	0.222	1.984	0.371
20.0	100.0	429.671	1.719	0.241	2.148	0.403

Table 7. (continued) Components of Standard Uncertainty in the Loss Modulus G"							
Temperature	Frequency	Loss Modulus G"	Standard Uncertainty (Type A)	$u(G",T)$	$u(G",\Omega)$	$u(G",\mathbf{M})$	$u(G",\Phi)$
°C	rad/s	Pa	Pa	Pa	Pa	Pa	Pa
30.0	0.03981	3.028	9.3×10^{-3}	0.012	3.0×10^{-4}	8.1×10^{-3}	7.6×10^{-3}
30.0	0.06310	4.773	0.011	0.019	4.7×10^{-4}	0.012	0.012
30.0	0.1000	7.480	0.016	0.030	7.3×10^{-4}	0.017	0.019
30.0	0.1585	11.650	0.023	0.045	1.1×10^{-3}	0.025	0.029
30.0	0.2512	17.936	0.032	0.067	1.6×10^{-3}	0.038	0.045
30.0	0.3981	27.172	0.055	0.095	2.3×10^{-3}	0.057	0.069
30.0	0.6310	40.190	0.077	0.131	3.2×10^{-3}	0.085	0.103
30.0	1.000	57.922	0.099	0.172	4.3×10^{-3}	0.124	0.152
30.0	1.585	80.873	0.133	0.217	5.5×10^{-3}	0.176	0.218
30.0	2.512	109.231	0.163	0.261	6.6×10^{-3}	0.246	0.305
30.0	3.981	142.586	0.209	0.297	7.7×10^{-3}	0.334	0.415
30.0	6.310	180.008	0.249	0.321	8.4×10^{-3}	0.443	0.552
30.0	10.00	219.989	0.298	0.327	8.8×10^{-3}	0.574	0.715
30.0	15.85	261.329	0.357	0.314	8.8×10^{-3}	0.726	0.905
30.0	25.12	302.326	0.424	0.284	8.3×10^{-3}	0.900	1.122
30.0	39.81	342.115	0.489	0.240	7.5×10^{-3}	1.092	1.363
30.0	63.10	380.371	0.571	0.191	6.6×10^{-3}	1.302	1.626
30.0	100.0	417.237	0.675	0.144	5.7×10^{-3}	1.527	1.906

Table 7. (continued) Components of Standard Uncertainty in the Loss Modulus G"						
Temperature	Frequency	Loss Modulus G"	$u(G'',h)$	$u(G'',R)$	Tilt	Evaporation
°C	rad/s	Pa	Pa	Pa	Pa	Pa
30.0	0.03981	3.028	0.012	1.7×10^{-3}	0.015	4.7×10^{-3}
30.0	0.06310	4.773	0.019	2.7×10^{-3}	0.024	8.5×10^{-3}
30.0	0.1000	7.480	0.030	4.2×10^{-3}	0.037	0.014
30.0	0.1585	11.650	0.047	6.5×10^{-3}	0.058	0.024
30.0	0.2512	17.936	0.072	0.010	0.090	0.037
30.0	0.3981	27.172	0.109	0.015	0.136	0.058
30.0	0.6310	40.190	0.161	0.023	0.201	0.086
30.0	1.000	57.922	0.232	0.032	0.290	0.125
30.0	1.585	80.873	0.323	0.045	0.404	0.176
30.0	2.512	109.231	0.437	0.061	0.546	0.238
30.0	3.981	142.586	0.570	0.080	0.713	0.312
30.0	6.310	180.008	0.720	0.101	0.900	0.395
30.0	10.00	219.989	0.880	0.123	1.100	0.484
30.0	15.85	261.329	1.045	0.146	1.307	0.576
30.0	25.12	302.326	1.209	0.169	1.512	0.668
30.0	39.81	342.115	1.368	0.192	1.711	0.758
30.0	63.10	380.371	1.521	0.213	1.902	0.845
30.0	100.0	417.237	1.669	0.234	2.086	0.928

Table 7. (continued) Components of Standard Uncertainty in the Loss Modulus G"							
Temperature	Frequency	Loss Modulus G"	Standard Uncertainty (Type A)	$u(G",T)$	$u(G",\Omega)$	$u(G",\mathbf{M})$	$u(G",\Phi)$
°C	rad/s	Pa	Pa	Pa	Pa	Pa	Pa
40.0	0.03981	2.047	6.7×10^{-3}	7.4×10^{-3}	2.0×10^{-4}	6.1×10^{-3}	5.1×10^{-3}
40.0	0.06310	3.236	0.011	0.012	3.2×10^{-4}	8.5×10^{-3}	8.1×10^{-3}
40.0	0.1000	5.101	8.0×10^{-3}	0.019	5.0×10^{-4}	0.012	0.013
40.0	0.1585	7.998	0.011	0.029	7.8×10^{-4}	0.018	0.020
40.0	0.2512	12.447	0.021	0.044	1.2×10^{-3}	0.027	0.031
40.0	0.3981	19.162	0.030	0.064	1.7×10^{-3}	0.041	0.048
40.0	0.6310	28.944	0.041	0.091	2.5×10^{-3}	0.061	0.074
40.0	1.000	42.757	0.061	0.125	3.4×10^{-3}	0.090	0.110
40.0	1.585	61.428	0.083	0.163	4.5×10^{-3}	0.131	0.161
40.0	2.512	85.515	0.108	0.205	5.7×10^{-3}	0.187	0.231
40.0	3.981	115.104	0.138	0.244	6.9×10^{-3}	0.260	0.322
40.0	6.310	149.769	0.183	0.277	7.9×10^{-3}	0.353	0.438
40.0	10.00	188.496	0.234	0.297	8.7×10^{-3}	0.467	0.581
40.0	15.85	229.749	0.290	0.301	9.1×10^{-3}	0.602	0.750
40.0	25.12	272.239	0.361	0.286	9.0×10^{-3}	0.761	0.948
40.0	39.81	314.635	0.442	0.256	8.5×10^{-3}	0.940	1.172
40.0	63.10	356.105	0.532	0.214	7.7×10^{-3}	1.139	1.421
40.0	100.0	396.285	0.670	0.167	6.7×10^{-3}	1.355	1.691

Table 7. (continued) Components of Standard Uncertainty in the Loss Modulus G"						
Temperature	Frequency	Loss Modulus G"	$u(G",h)$	$u(G",R)$	Tilt	Evaporation
°C	rad/s	Pa	Pa	Pa	Pa	Pa
40.0	0.03981	2.047	8.2×10^{-3}	1.2×10^{-3}	0.010	7.2×10^{-3}
40.0	0.06310	3.236	0.013	1.8×10^{-3}	0.016	0.013
40.0	0.1000	5.101	0.020	2.9×10^{-3}	0.026	0.022
40.0	0.1585	7.998	0.032	4.5×10^{-3}	0.040	0.036
40.0	0.2512	12.447	0.050	7.0×10^{-3}	0.062	0.058
40.0	0.3981	19.162	0.077	0.011	0.096	0.091
40.0	0.6310	28.944	0.116	0.016	0.145	0.140
40.0	1.000	42.757	0.171	0.024	0.214	0.208
40.0	1.585	61.428	0.246	0.034	0.307	0.300
40.0	2.512	85.515	0.342	0.048	0.428	0.419
40.0	3.981	115.104	0.460	0.064	0.576	0.566
40.0	6.310	149.769	0.599	0.084	0.749	0.738
40.0	10.00	188.496	0.754	0.106	0.942	0.932
40.0	15.85	229.749	0.919	0.129	1.149	1.139
40.0	25.12	272.239	1.089	0.152	1.361	1.352
40.0	39.81	314.635	1.259	0.176	1.573	1.567
40.0	63.10	356.105	1.424	0.199	1.781	1.777
40.0	100.0	396.285	1.585	0.222	1.981	1.981

Table 7. (continued) Components of Standard Uncertainty in the Loss Modulus G"							
Temperature	Frequency	Loss Modulus G"	Standard Uncertainty (Type A)	$u(G",T)$	$u(G",\Omega)$	$u(G",M)$	$u(G",\Phi)$
°C	rad/s	Pa	Pa	Pa	Pa	Pa	Pa
50.0	0.03981	1.457	4.8×10^{-3}	4.6×10^{-3}	1.4×10^{-4}	5.0×10^{-3}	3.6×10^{-3}
50.0	0.06310	2.316	7.9×10^{-3}	7.5×10^{-3}	2.2×10^{-4}	6.7×10^{-3}	5.8×10^{-3}
50.0	0.1000	3.630	8.7×10^{-3}	0.012	3.6×10^{-4}	9.3×10^{-3}	9.1×10^{-3}
50.0	0.1585	5.726	0.014	0.019	5.6×10^{-4}	0.013	0.014
50.0	0.2512	8.980	0.018	0.029	8.6×10^{-4}	0.020	0.022
50.0	0.3981	13.949	0.028	0.044	1.3×10^{-3}	0.030	0.035
50.0	0.6310	21.392	0.040	0.064	1.9×10^{-3}	0.045	0.054
50.0	1.000	32.210	0.057	0.090	2.7×10^{-3}	0.068	0.082
50.0	1.585	47.339	0.081	0.122	3.7×10^{-3}	0.100	0.122
50.0	2.512	67.610	0.108	0.159	4.8×10^{-3}	0.145	0.179
50.0	3.981	93.473	0.140	0.197	6.1×10^{-3}	0.205	0.254
50.0	6.310	125.025	0.177	0.233	7.3×10^{-3}	0.284	0.353
50.0	10.00	161.601	0.223	0.261	8.3×10^{-3}	0.384	0.477
50.0	15.85	201.983	0.270	0.276	9.0×10^{-3}	0.505	0.629
50.0	25.12	244.960	0.326	0.276	9.3×10^{-3}	0.649	0.809
50.0	39.81	289.041	0.389	0.260	9.1×10^{-3}	0.816	1.018
50.0	63.10	333.071	0.465	0.228	8.6×10^{-3}	1.004	1.253
50.0	100.0	376.368	0.572	0.187	7.7×10^{-3}	1.212	1.512

Table 7. (continued) Components of Standard Uncertainty in the Loss Modulus G"						
Temperature	Frequency	Loss Modulus G"	$u(G'',h)$	$u(G'',R)$	Tilt	Evaporation
°C	rad/s	Pa	Pa	Pa	Pa	Pa
50.0	0.03981	1.457	5.8×10^{-3}	8.2×10^{-4}	7.3×10^{-3}	0.011
50.0	0.06310	2.316	9.3×10^{-3}	1.3×10^{-3}	0.012	0.020
50.0	0.1000	3.630	0.015	2.0×10^{-3}	0.018	0.034
50.0	0.1585	5.726	0.023	3.2×10^{-3}	0.029	0.056
50.0	0.2512	8.980	0.036	5.0×10^{-3}	0.045	0.091
50.0	0.3981	13.949	0.056	7.8×10^{-3}	0.070	0.143
50.0	0.6310	21.392	0.086	0.012	0.107	0.223
50.0	1.000	32.210	0.129	0.018	0.161	0.338
50.0	1.585	47.339	0.189	0.027	0.237	0.499
50.0	2.512	67.610	0.270	0.038	0.338	0.715
50.0	3.981	93.473	0.374	0.052	0.467	0.991
50.0	6.310	125.025	0.500	0.070	0.625	1.328
50.0	10.00	161.601	0.646	0.090	0.808	1.721
50.0	15.85	201.983	0.808	0.113	1.010	2.157
50.0	25.12	244.960	0.980	0.137	1.225	2.621
50.0	39.81	289.041	1.156	0.162	1.445	3.101
50.0	63.10	333.071	1.332	0.187	1.665	3.580
50.0	100.0	376.368	1.505	0.211	1.882	4.052

5. Conclusion

The measured viscosity and first normal stress difference data are given in Table 8, along with the combined standard uncertainties and the models fit to the data using equations (6) through (8).

The measured storage modulus G' and loss modulus G" are given in Table 9, along with the combined standard uncertainties and models fit to the data using equations (39) and (40).

Table 8. Viscosity and First Normal Stress Difference (N_1) with Combined Standard Uncertainties and Models Fit to the Data							
Temperature	Shear Rate	Viscosity	Combined Standard Uncertainty	Model Fit to Viscosity	N_1	Combined Standard Uncertainty	Model Fit to N_1
°C	s^{-1}	Pa·s	Pa·s	Pa·s	Pa	Pa	Pa
0.0	0.001000	382.90	5.97	393.97			
0.0	0.001585	380.05	5.27	393.30			
0.0	0.002512	382.38	4.93	392.33			
0.0	0.003981	382.90	4.74	390.93			
0.0	0.006310	383.97	4.62	388.91			
0.0	0.01000	383.08	4.52	386.04			
0.0	0.01585	382.90	4.42	381.95			
0.0	0.02512	379.83	4.31	376.18			
0.0	0.03981	375.06	4.15	368.12			
0.0	0.06310	365.51	3.95	357.03			
0.0	0.1000	350.01	3.66	342.11			
0.0	0.1585	328.43	3.30	322.58	16.30	5.49	24.88
0.0	0.2512	300.79	2.87	297.94	43.78	4.84	46.98
0.0	0.3981	268.52	2.39	268.25	91.85	5.64	83.99
0.0	0.6310	232.98	1.90	234.42	149.06	6.93	143.09
0.0	1.000	196.53	1.44	198.21	257.45	7.41	234.03
0.0	1.585	161.10	1.04	161.96	372.45	7.41	369.92
0.0	2.512	128.41	0.73	128.04	572.64	8.43	567.68
0.0	3.981	99.45	0.49	98.23	845.17	10.41	847.95
0.0	6.310	75.07	0.33	73.46	1218.81	14.41	1234.06
0.0	10.00	55.59	0.22	53.81	1717.43	17.68	1750.11
0.0	15.85	40.26	0.15	38.78	2362.56	22.19	2418.51
0.0	25.12	28.67	0.10	27.60	3195.73	27.63	3258.06
0.0	39.81	20.23	0.08	19.48	4251.21	34.35	4283.30
0.0	63.10	13.88	0.08	13.65	5518.63	43.41	5505.89
0.0	100.0	9.08	0.06	9.53	7125.98	53.74	6937.36

Table 8. (continued) Viscosity and First Normal Stress Difference (N_1) with Combined Standard Uncertainties and Models Fit to the Data

Temperature	Shear Rate	Viscosity	Combined Standard Uncertainty	Model Fit to Viscosity	N_1	Combined Standard Uncertainty	Model Fit to N_1
°C	s^{-1}	Pa·s	Pa·s	Pa·s	Pa	Pa	Pa
25.0	0.001000	97.93	3.52	100.09			
25.0	0.001585	98.05	2.46	100.03			
25.0	0.002512	98.25	1.86	99.96			
25.0	0.003981	97.93	1.45	99.84			
25.0	0.006310	98.35	1.26	99.68			
25.0	0.01000	98.06	1.17	99.45			
25.0	0.01585	98.72	1.12	99.11			
25.0	0.02512	98.78	1.09	98.63			
25.0	0.03981	98.61	1.07	97.93			
25.0	0.06310	98.37	1.05	96.95			
25.0	0.1000	97.52	1.02	95.55			
25.0	0.1585	96.13	0.99	93.60	2.35	0.95	2.55
25.0	0.2512	93.71	0.94	90.91	5.51	0.89	5.69
25.0	0.3981	90.00	0.88	87.28	12.92	0.94	12.10
25.0	0.6310	84.59	0.80	82.50	26.54	1.05	24.39
25.0	1.000	77.61	0.70	76.44	50.05	1.24	46.41
25.0	1.585	69.16	0.58	69.09	87.62	1.64	83.57
25.0	2.512	59.98	0.47	60.64	148.18	2.31	143.27
25.0	3.981	50.56	0.36	51.51	236.76	3.72	235.55
25.0	6.310	41.44	0.27	42.29	377.26	6.07	373.98
25.0	10.00	33.04	0.19	33.58	585.07	8.10	576.17
25.0	15.85	25.61	0.14	25.87	880.00	10.41	863.78
25.0	25.12	19.36	0.10	19.41	1280.30	13.63	1261.63
25.0	39.81	14.26	0.08	14.25	1800.04	17.78	1795.66
25.0	63.10	10.22	0.07	10.29	2462.22	25.58	2490.38
25.0	100.0	7.22	0.06	7.34	3318.59	31.68	3366.61

Table 8. (continued) Viscosity and First Normal Stress Difference (N_1) with Combined Standard Uncertainties and Models Fit to the Data							
Temperature	Shear Rate	Viscosity	Combined Standard Uncertainty	Model Fit to Viscosity	N_1	Combined Standard Uncertainty	Model Fit to N_1
°C	s^{-1}	Pa·s	Pa·s	Pa·s	Pa	Pa	Pa
50.0	0.001000	36.74	3.25	38.07			
50.0	0.001585	37.20	2.11	38.06			
50.0	0.002512	37.62	1.40	38.05			
50.0	0.003981	37.30	0.98	38.03			
50.0	0.006310	37.72	0.74	38.00			
50.0	0.01000	37.73	0.63	37.97			
50.0	0.01585	37.50	0.57	37.91			
50.0	0.02512	37.55	0.55	37.83			
50.0	0.03981	37.83	0.54	37.71			
50.0	0.06310	37.80	0.54	37.54			
50.0	0.1000	37.76	0.55	37.30			
50.0	0.1585	37.66	0.55	36.96			
50.0	0.2512	37.39	0.55	36.47			
50.0	0.3981	36.93	0.55	35.78	2.38	0.84	2.23
50.0	0.6310	36.10	0.54	34.84	5.08	0.86	5.02
50.0	1.000	34.77	0.52	33.55	10.73	0.91	10.81
50.0	1.585	32.79	0.50	31.84	21.47	1.00	22.09
50.0	2.512	30.20	0.46	29.66	41.05	1.33	42.62
50.0	3.981	27.05	0.42	26.98	68.85	2.14	77.74
50.0	6.310	23.54	0.37	23.86	114.48	3.82	134.77
50.0	10.00	19.95	0.32	20.43	202.50	5.71	223.69
50.0	15.85	16.41	0.27	16.92	344.80	7.85	357.98
50.0	25.12	13.14	0.22	13.54	557.75	11.34	555.37
50.0	39.81	10.24	0.17	10.50	860.20	16.37	837.99
50.0	63.10	7.75	0.13	7.93	1269.10	23.53	1231.67
50.0	100.0	5.72	0.10	5.85	1796.79	33.56	1764.04

Table 9. Storage Modulus G' and Loss Modulus G" with the Combined Standard Uncertainties and Models Fit to the Data							
Temperature	Frequency	Storage Modulus G'	Combined Standard Uncertainty in G'	Model Fit to G'	Loss Modulus G"	Combined Standard Uncertainty in G"	Model Fit to G"
°C	rad/s	Pa	Pa	Pa	Pa	Pa	Pa
0.0	0.02512	0.605	0.029	0.565	7.976	0.076	8.064
0.0	0.03981	1.385	0.045	1.305	12.346	0.116	12.519
0.0	0.06310	2.973	0.071	2.845	18.852	0.175	19.124
0.0	0.1000	6.067	0.114	5.849	28.201	0.256	28.609
0.0	0.1585	11.699	0.184	11.344	41.113	0.366	41.753
0.0	0.2512	21.207	0.289	20.762	58.262	0.510	59.264
0.0	0.3981	36.214	0.440	35.897	79.846	0.690	81.618
0.0	0.6310	58.511	0.652	58.727	105.855	0.912	108.890
0.0	1.000	89.636	0.929	91.103	135.566	1.150	140.620
0.0	1.585	130.948	1.270	134.379	168.058	1.431	175.769
0.0	2.512	182.959	1.690	189.081	201.883	1.727	212.800
0.0	3.981	246.085	2.180	254.775	235.898	2.045	249.910
0.0	6.310	319.791	2.752	330.190	268.794	2.381	285.327
0.0	10.00	403.264	3.381	413.644	299.722	2.718	317.654
0.0	15.85	495.279	4.070	503.665	328.134	3.065	346.137
0.0	25.12	593.784	4.832	599.721	354.144	3.417	370.841
0.0	39.81	697.210	5.646	702.968	378.885	3.778	392.707
0.0	63.10	804.155	6.490	817.012	404.134	4.163	413.539
0.0	100.0	912.794	7.392	948.855	432.131	4.589	436.001

Table 9. (continued) Storage Modulus G' and Loss Modulus G" with the Combined Standard Uncertainties and Models Fit to the Data							
Temperature	Frequency	Storage Modulus G'	Combined Standard Uncertainty in G'	Model Fit to G'	Loss Modulus G"	Combined Standard Uncertainty in G"	Model Fit to G"
°C	rad/s	Pa	Pa	Pa	Pa	Pa	Pa
10.0	0.03981	0.486	0.026	0.469	7.295	0.067	7.438
10.0	0.06310	1.099	0.041	1.099	11.352	0.104	11.585
10.0	0.1000	2.438	0.064	2.430	17.419	0.157	17.776
10.0	0.1585	5.076	0.102	5.070	26.287	0.233	26.741
10.0	0.2512	10.023	0.164	9.976	38.734	0.338	39.279
10.0	0.3981	18.531	0.258	18.520	55.454	0.476	56.151
10.0	0.6310	32.329	0.399	32.471	76.894	0.652	77.925
10.0	1.000	53.217	0.598	53.844	103.161	0.870	104.795
10.0	1.585	83.154	0.861	84.616	133.741	1.119	136.432
10.0	2.512	123.313	1.200	126.341	167.622	1.406	171.910
10.0	3.981	174.767	1.605	179.799	203.424	1.712	209.757
10.0	6.310	238.200	2.095	244.784	240.105	2.044	248.155
10.0	10.00	313.204	2.668	320.173	276.093	2.397	285.239
10.0	15.85	398.816	3.325	404.283	310.338	2.763	319.442
10.0	25.12	493.868	4.045	495.473	342.151	3.133	349.801
10.0	39.81	596.352	4.831	592.887	371.952	3.512	376.167
10.0	63.10	704.450	5.670	697.213	400.674	3.910	399.285
10.0	100.0	816.167	6.549	811.461	429.896	4.336	420.795

Table 9. (continued) Storage Modulus G' and Loss Modulus G" with the Combined Standard Uncertainties and Models Fit to the Data							
Temper-ature	Frequency	Storage Modulus G'	Combined Standard Uncertainty in G'	Model Fit to G'	Loss Modulus G"	Combined Standard Uncertainty in G"	Model Fit to G"
°C	rad/s	Pa	Pa	Pa	Pa	Pa	Pa
20.0	0.03981	0.182	0.017	0.175	4.605	0.041	4.622
20.0	0.06310	0.430	0.027	0.437	7.259	0.064	7.272
20.0	0.1000	1.023	0.040	1.029	11.277	0.100	11.341
20.0	0.1585	2.275	0.063	2.291	17.376	0.151	17.437
20.0	0.2512	4.809	0.100	4.812	26.304	0.227	26.296
20.0	0.3981	9.571	0.160	9.530	38.913	0.329	38.738
20.0	0.6310	17.918	0.252	17.811	56.012	0.470	55.557
20.0	1.000	31.620	0.389	31.430	78.152	0.646	77.370
20.0	1.585	52.610	0.584	52.447	105.458	0.865	104.431
20.0	2.512	82.735	0.845	82.918	137.487	1.123	136.466
20.0	3.981	123.780	1.185	124.513	173.320	1.417	172.594
20.0	6.310	176.784	1.607	178.141	211.561	1.740	211.357
20.0	10.00	242.540	2.107	243.709	250.993	2.088	250.910
20.0	15.85	320.640	2.693	320.155	290.060	2.456	289.320
20.0	25.12	410.368	3.364	405.783	327.457	2.837	324.922
20.0	39.81	510.091	4.112	498.867	362.955	3.233	356.644
20.0	63.10	617.780	4.919	598.391	396.760	3.643	384.228
20.0	100.0	731.238	5.776	704.841	429.671	4.077	408.333

Table 9. (continued) Storage Modulus G' and Loss Modulus G" with the Combined Standard Uncertainties and Models Fit to the Data							
Temperature	Frequency	Storage Modulus G'	Combined Standard Uncertainty in G'	Model Fit to G'	Loss Modulus G"	Combined Standard Uncertainty in G"	Model Fit to G"
°C	rad/s	Pa	Pa	Pa	Pa	Pa	Pa
30.0	0.03981	0.077	0.011	0.068	3.028	0.028	3.011
30.0	0.06310	0.178	0.018	0.180	4.773	0.042	4.748
30.0	0.1000	0.440	0.027	0.448	7.480	0.066	7.471
30.0	0.1585	1.039	0.041	1.057	11.650	0.101	11.652
30.0	0.2512	2.323	0.065	2.353	17.936	0.154	17.914
30.0	0.3981	4.899	0.103	4.942	27.172	0.232	27.017
30.0	0.6310	9.776	0.164	9.789	40.190	0.339	39.801
30.0	1.000	18.376	0.259	18.295	57.922	0.482	57.084
30.0	1.585	32.459	0.401	32.287	80.873	0.667	79.500
30.0	2.512	54.041	0.601	53.879	109.231	0.893	107.309
30.0	3.981	85.240	0.872	85.188	142.586	1.163	140.234
30.0	6.310	127.700	1.228	127.930	180.008	1.470	177.366
30.0	10.00	182.644	1.667	183.040	219.989	1.808	217.209
30.0	15.85	250.728	2.198	250.425	261.329	2.174	257.867
30.0	25.12	331.736	2.820	328.992	302.326	2.559	297.352
30.0	39.81	424.451	3.521	417.002	342.115	2.959	333.953
30.0	63.10	527.345	4.299	512.676	380.371	3.375	366.566
30.0	100.0	638.186	5.139	614.971	417.237	3.807	394.925

Table 9. (continued) Storage Modulus G' and Loss Modulus G" with the Combined Standard Uncertainties and Models Fit to the Data							
Temper-ature	Frequency	Storage Modulus G'	Combined Standard Uncertainty in G'	Model Fit to G'	Loss Modulus G"	Combined Standard Uncertainty in G"	Model Fit to G"
°C	rad/s	Pa	Pa	Pa	Pa	Pa	Pa
40.0	0.03981	0.030	0.0094	0.028	2.047	0.020	2.056
40.0	0.06310	0.062	0.013	0.077	3.236	0.032	3.229
40.0	0.1000	0.200	0.019	0.202	5.101	0.048	5.092
40.0	0.1585	0.478	0.028	0.501	7.998	0.075	8.007
40.0	0.2512	1.142	0.044	1.174	12.447	0.118	12.474
40.0	0.3981	2.562	0.070	2.600	19.162	0.180	19.147
40.0	0.6310	5.391	0.111	5.429	28.944	0.271	28.815
40.0	1.000	10.695	0.179	10.695	42.757	0.398	42.348
40.0	1.585	19.966	0.285	19.879	61.428	0.568	60.574
40.0	2.512	35.035	0.445	34.894	85.515	0.787	84.114
40.0	3.981	58.033	0.676	57.927	115.104	1.057	113.191
40.0	6.310	91.010	0.992	91.127	149.769	1.378	147.460
40.0	10.00	135.717	1.407	136.200	188.496	1.742	185.926
40.0	15.85	193.163	1.929	194.010	229.749	2.140	227.005
40.0	25.12	264.093	2.562	264.356	272.239	2.565	268.722
40.0	39.81	347.741	3.302	346.035	314.635	3.011	309.050
40.0	63.10	443.102	4.141	437.224	356.105	3.470	346.277
40.0	100.0	548.169	5.067	536.136	396.285	3.948	379.343

Table 9. (continued) Storage Modulus G' and Loss Modulus G" with the Combined Standard Uncertainties and Models Fit to the Data							
Temperature	Frequency	Storage Modulus G'	Combined Standard Uncertainty in G'	Model Fit to G'	Loss Modulus G"	Combined Standard Uncertainty in G"	Model Fit to G"
°C	rad/s	Pa	Pa	Pa	Pa	Pa	Pa
50.0	0.03981	0.013	0.0083	0.012	1.457	0.017	1.468
50.0	0.06310	0.034	0.011	0.035	2.316	0.029	2.284
50.0	0.1000	0.088	0.014	0.094	3.630	0.046	3.592
50.0	0.1585	0.241	0.020	0.245	5.726	0.074	5.665
50.0	0.2512	0.592	0.032	0.601	8.980	0.117	8.897
50.0	0.3981	1.375	0.051	1.394	13.949	0.183	13.830
50.0	0.6310	3.058	0.083	3.055	21.392	0.281	21.160
50.0	1.000	6.385	0.139	6.313	32.210	0.424	31.720
50.0	1.585	12.509	0.232	12.308	47.339	0.623	46.402
50.0	2.512	23.028	0.385	22.641	67.610	0.888	66.032
50.0	3.981	39.969	0.622	39.343	93.473	1.228	91.190
50.0	6.310	65.434	0.968	64.677	125.025	1.643	122.010
50.0	10.00	101.548	1.451	100.798	161.601	2.129	158.023
50.0	15.85	149.841	2.091	149.329	201.983	2.671	198.091
50.0	25.12	211.567	2.901	210.969	244.960	3.257	240.496
50.0	39.81	286.640	3.882	285.316	289.041	3.871	283.179
50.0	63.10	374.506	5.029	370.982	333.071	4.499	324.088
50.0	100.0	473.567	6.320	466.051	376.368	5.137	361.564

6. References

1. K. Walters, *Rheometry*, Chapman and Hall, London, 1975.
2. W.R. Schowalter, *Mechanics of Non-Newtonian Fluids*, Pergamon Press, Oxford, 1978.
3. R.I. Tanner, *Engineering Rheology*, Clarendon Press, Oxford, 1985.
4. R.B. Bird, R.C. Armstrong and O. Hassager, *Dynamics of Polymeric Liquids, Vol. 1, Fluid Mechanics*, 2nd Edition, John Wiley and Sons, New York, 1987.
5. R.B. Bird, C.F. Curtiss, R.C. Armstrong and O. Hassager, *Dynamics of Polymeric Liquids, Volume 2, Kinetic Theory*, 2nd Edition, John Wiley and Sons, New York, 1987.
6. C.W. Macosko, *Rheology - Principles, Measurements and Applications*, Wiley-VCH, New York, 1994.
7. L.J. Zapas, and J.C. Phillips, *Certificate for Standard Reference Material 1490, Polyisobutylene Solution in Cetane (Viscosity and First Normal Stress Difference)*, National Bureau of Standards, Office of Standard Reference Materials, 1977.
8. *CRC Handbook of Chemistry and Physics*, 81st Edition, CRC Press, Boca Raton, 2000.
9. L.J. Zapas and J.C. Phillips, "Simple Shearing Flows in Polyisobutylene Solutions," *Journal of Research of the National Bureau of Standards - A. Physics and Chemistry*, Volume 75A, pp. 33-40, 1971.
10. L.J. Zapas and J.C. Phillips, "Nonlinear Behavior of Polyisobutylene Solutions as a Function of Concentration," *Journal of Rheology*, Volume 25, pp. 405-420, 1981.
11. K. Khalil, A. Tougui and D. Sigli, "Relation between some rheological properties of polyisobutylene solutions and their mode of preparation," *Journal of Non-Newtonian Fluid Mechanics*, Volume 52, pp. 375-386, 1994.
12. C.R. Schultheisz and G.B. McKenna, "A Nonlinear Fluid Standard Reference Material: Progress Report," *Proceedings of SPE ANTEC*, Volume 1, Processing, pp. 1125-1129, 1999.
13. C.R. Schultheisz and G.B. McKenna, "Standard Reference Materials: Non-Newtonian Fluids for Rheological Measurements," *Proceedings of SPE ANTEC*, Volume 1, Processing, pp. 1042-1046, 2000.
14. *Merck Index*, Twelfth Edition, Merck and Company, Rahway, New Jersey, 1996.
15. *Webster's Third New International Dictionary of the English Language*, Merriam-Webster, Inc., Springfield, Massachusetts, 1981.
16. Advanced Rheometric Expansion System ARES Instrument Manual, Rheometric Scientific, 1997.
17. J.V. Nicholas and D.R. White, *Traceable Temperatures*, John Wiley and Sons, Chichester, 1994.
18. B.N. Taylor, and C.E. Kuyatt, "Guidelines for Evaluating and Expressing the Uncertainty of NIST Measurement Results," NIST Technical Note 1297, 1994 Edition, U.S. Government Printing Office, Washington, D.C., 1994.
19. M.E. Mackay and G.K. Dick, "A comparison of the dimensions of a truncated cone measured with various techniques: The cone measurement project," *Journal of Rheology*, Volume 39, pp. 673-677, 1995.
20. R. Garritano, Rheometric Scientific, Incorporated, private communication.
21. J.M. Niemiec, J.-J. Pesce, G.B. McKenna, S. Skocypec and R.F. Garritano, "Anomalies in the Normal Force Measurement when using a Force Rebalance Transducer," *Journal of Rheology*, Volume 40, pp. 323-334, 1996.

22. Schultheisz, C.R. and McKenna, G.B., "Thermal Expansion and Its Effect on the Normal Force for a Modified Force Rebalance Transducer," *Proceedings of SPE ANTEC*, Volume 1, Processing, pp. 1142-1145, May 3-6, 1999.
23. H. Markovitz, L.J. Elyash, F.J. Padden, Jr. and T.W. DeWitt, "A Cone-and-Plate Viscometer," *Journal of Colloid Science*, Volume 10, p. 165-173, 1955.
24. W.H. Press, S.A. Teukolsky, W.T. Vetterling and B.P. Flannery, *Numerical Recipes*, Second Edition, Cambridge University Press, Cambridge, 1992.
25. G.V. Gordon and M.T. Shaw, *Computer Programs for Rheologists*, Hanser Publishers, Munich, 1994.
26. D.W. Van Krevelen, *Properties of Polymers*, Third Edition, Elsevier, Amsterdam, 1990.
27. W.J. Lyman, W.F. Reehl and D.H. Rosenblatt, *Handbook of Chemical Property Estimation Methods*, American Chemical Society, Washington, D.C., 1990.
28. B.D. Marsh and J.R.A. Pearson, "The Measurement of Normal-Stress Differences Using a Cone-and-Plate Total Thrust Apparatus," *Rheologica Acta*, Volume 7, pp. 326-331, 1968.
29. C.L. Yaws, *Handbook of Thermal Conductivity*, Gulf Publishing Company, Houston, 1995.
30. G. Winther, D.M. Parsons and J.L. Schrag, "A High-Speed, High-Precision Data Acquisition and Processing System for Experiments Producing Steady State Periodic Signals," *Journal of Polymer Science: Part B: Polymer Physics*, Volume 32, pp. 659-670, 1994.
31. *RMS-800/RDS-II Instrument Manual*, Rheometric Scientific, 1990.

National Institute of Standards & Technology

Certificate of Analysis

Standard Reference Material® 2490

Non-Newtonian Polymer Solution for Rheology

Polyisobutylene Dissolved in 2,6,10,14-Tetramethylpentadecane

This Standard Reference Material (SRM) is intended primarily for use in calibration and performance evaluation of instruments used to determine the viscosity and first normal stress difference in steady shear, or to determine the dynamic mechanical storage and loss moduli and shift factors through time-temperature superposition. SRM 2490 consists of a polyisobutylene dissolved in 2,6,10,14-tetramethylpentadecane (common name pristane). The solution contains a mass fraction of 0.114 polyisobutylene. The mass average relative molecular mass of the polyisobutylene is reported as 1 000 000 by the supplier. One unit of SRM 2490 consists of 100 mL of the solution packaged in an amber glass bottle.

Certified Values and Uncertainties: The certified values of the viscosity and first normal stress difference as functions of shear rate are given in Tables 4a, 4b, and 4c at temperatures of 0 °C, 25 °C, and 50 °C, respectively. Tables 4a through 4c also list the expanded combined uncertainties in the certified values of the viscosity and first normal stress difference. Tables 5a, 5b, 5c, 5d, 5e, and 5f list the certified values of the storage modulus G' and loss modulus G" as functions of frequency at 0 °C, 10 °C, 20 °C, 30 °C, 40 °C, and 50 °C, respectively. Tables 5a through 5f also list the expanded combined uncertainties in the certified values of the storage modulus G' and loss modulus G". The uncertainties in Tables 4a through 4c and 5a through 5f were calculated as $U = ku_c$, where $k = 2$ is the coverage factor for a 95 % level of confidence and u_c is the combined standard uncertainty calculated according to the ISO Guide [1].

Expiration of Certification: The certification of SRM 2490 is valid until **31 December 2008**, within the measurement uncertainties specified, provided that the SRM is handled in accordance with the storage instructions given in this certificate. This certification is nullified if the SRM is modified or contaminated.

Maintenance of SRM Certification: NIST will monitor this SRM over the period of its certification. If substantive technical changes occur that affect the certification before expiration of this certificate, NIST will notify the purchaser. Return of the attached registration card will facilitate notification.

Technical coordination leading to the certification of this SRM was provided by B.M. Fanconi of the NIST Polymers Division.

The certification of this SRM was performed by C.R. Schultheisz of the NIST Polymers Division.

The support aspects in the preparation, certification, and issuance of this SRM were coordinated through the NIST Standard Reference Materials Program by J.W.L. Thomas.

Eric J. Amis, Chief
Polymers Division

Gaithersburg, MD 20899
Certificate Issue Date: 18 September 2001

John Rumble, Jr., Acting Chief
Standard Reference Materials Program

Statistical analysis and measurement advice were provided by S.D. Leigh of the NIST Statistical Engineering Division.

Technical assistance and advice were provided by Gregory Strouse and Dawn Vaughn of the NIST Process Measurements Division and Gregory B. McKenna of the Texas Tech University.

Source of Material: The polyisobutylene and 2,6,10,14-tetramethylpentadecane were obtained from Aldrich Chemical Company, Milwaukee, Wisconsin.[1] The solution was mixed and packaged by the Cannon Instrument Company, State College, PA.[1]

Storage and Handling: The SRM should be stored in the original bottle with the lid tightly closed under normal laboratory conditions. Before taking a sample, the bottle should be turned end-over-end at a rate of approximately 1 revolution per 10 minutes for 30 minutes. This procedure is intended to ensure that the material in each bottle is homogeneous, in case there is any settling caused by gravity.

Homogeneity and Characterization: The homogeneity of SRM 2490 was tested by measuring the zero-shear-rate viscosity at 25 °C from 10 bottles randomly chosen from the 438 bottles available. Three samples from each bottle were tested in random order. The characterization of this polymer solution is described in reference [2].

Measurement Technique: All rheological testing was carried out using a Rheometric Scientific, Inc., ARES controlled-strain rheometer.[1] Transducer calibration was accomplished, in accordance with the manufacturer's instructions, by hanging a known mass from a fixture mounted to the transducer to apply a known torque or normal force. Phase angle calibration was accomplished, also in accordance with the manufacturer's instructions, by applying an oscillatory strain to an elastic steel test coupon. Temperature calibration in the rheometer was accomplished through comparison with a NIST-calibrated thermistor. The viscosity and first normal stress difference were measured in steady shear using 50 mm diameter, 0.02 rad cone-and-plate fixtures. The storage modulus and loss modulus were measured in 50 mm diameter parallel-plate fixtures with an applied strain magnitude of 20 % at a nominal gap of 1 mm.

Models for the Data: The steady shear data (viscosity and first normal stress difference) and the oscillatory data (storage modulus and loss modulus) were fitted to empirical functions to describe master curves and calculate shift factors for time-temperature superposition. These models can be used to estimate the rheological behavior of the material in the temperature range 0 °C to 50 °C.

Models for the Steady Shear Data: The viscosity $\eta(\dot{\gamma},T)$ as a function of the shear rate $\dot{\gamma}$, and the temperature T was fitted to a Cross model [3,4] of the form

$$\eta(\dot{\gamma},T) = \left(\frac{T\rho}{T_R \rho_R}\right)\left(\frac{\eta_R a(T)}{1+(\xi_0 a(T)\dot{\gamma})^{1-n}}\right) \qquad (1)$$

where ρ is the density at temperature T, η_R is the zero-shear-rate viscosity at the reference temperature $T_R = 25$ °C, ρ_R is the density at the reference temperature T_R, ξ_0 is a parameter that governs the transition from the Newtonian regime at low shear rates to the power law regime at high shear rates, and n is the power at which the shear stress increases with shear rate. The density was approximated as a linear function of temperature, with $\rho(T) = \rho_R(1-\alpha(T-T_R))$, where $\alpha = 6 \times 10^{-4}$ cm^3/(cm^3 K) is the volumetric coefficient of thermal expansion. The shift factor $a(T)$ was fitted with a function of the WLF type [3],

$$a(T) = \exp\left(\frac{-C_1(T-T_R)}{C_2+T-T_R}\right) \qquad (2)$$

---

[1]Certain commercial equipment, instrumentation, or materials are identified in this certificate to specify adequately the experimental procedure. Such identification does not imply recommendation or endorsement by the NIST, nor does it imply that the materials or equipment identified are necessarily the best available for the purpose.

The parameters η_R, ξ_0, n, C_1, and C_2 estimated from the fit to the viscosity data are given in Table 1.

Table 1. Parameters for $\eta(\dot{\gamma},T)$ and $a(T)$

Parameter	Value	Standard Uncertainty
η_R	100.2 Pa·s	0.6 Pa·s
ξ_0	0.234 s	0.004 s
n	0.195	0.004
C_1	7.23	0.24
C_2	150 °C	5 °C

The first normal stress difference $N_1(\dot{\gamma},T)$ was fitted to a similar empirical model using the same temperature shift factor $a(T)$ calculated from the viscosity data:

$$N_1(\dot{\gamma},T) = \left(\frac{T\rho}{T_R \rho_R}\right)\left(\frac{\psi_R (a(T)\dot{\gamma})^2}{1 + \xi_1 a(T)\dot{\gamma} + (\xi_2 a(T)\dot{\gamma})^p}\right) \quad (3)$$

where ρ is the density at temperature T; ψ_R is the zero-shear-rate first normal stress coefficient at the reference temperature $T_R = 25$ °C; ρ_R is the density at the reference temperature T_R; and ξ_1, ξ_2, and p are parameters estimated from the fit to the data. The density was approximated as a linear function of temperature, with $\rho(T) = \rho_R(1 - \alpha(T - T_R))$, where $\alpha = 6 \times 10^{-4}$ cm³/(cm³ K). Values for the parameters describing $N_1(\dot{\gamma},T)$ are given in Table 2.

Table 2. Parameters for $N_1(\dot{\gamma},T)$

Parameter	Value	Standard Uncertainty
ψ_R	129 Pa·s²	5 Pa·s²
ξ_1	1.69 s	0.13 s
ξ_2	0.247 s	0.026 s
p	1.67	0.047

Models for the Oscillatory Data: The storage modulus $G'(\Omega,T)$ and loss modulus $G''(\Omega,T)$ as functions of the frequency of oscillation Ω and temperature T were modeled using polynomial functions [4]. The data were fitted to functions of the form

$$\ln\left(\frac{G'(\Omega,T)}{1 \text{ Pa}}\right) = \ln\left(\frac{T\rho}{T_R \rho_R}\right) + \sum_{k=0}^{4} p_k \left(\ln\left(\frac{a(T)\Omega}{1 \text{ rad/s}}\right)\right)^k$$

$$\ln\left(\frac{G''(\Omega,T)}{1 \text{ Pa}}\right) = \ln\left(\frac{T\rho}{T_R \rho_R}\right) + \sum_{k=0}^{4} q_k \left(\ln\left(\frac{a(T)\Omega}{1 \text{ rad/s}}\right)\right)^k \quad (4)$$

where ρ is the density at temperature T, and ρ_R is the density at the reference temperature $T_R = 25$ °C. The density was again approximated as a linear function of temperature, with $\rho(T) = \rho_R(1-\alpha(T-T_R))$, where $\alpha = 6 \times 10^{-4}$ cm^3/(cm^3 K). The shift factor $a(T)$ again was fitted with a function of the WLF type [3],

$$a(T) = \exp\left(\frac{-C_1(T-T_R)}{C_2 + T - T_R}\right) \quad (5)$$

The parameters estimated from the to the oscillatory data are given in Table 3.

Table 3. Parameters for G'(Ω, T), G''(Ω, T) and $a(T)$

Parameter	Value	Standard Uncertainty
p_0	3.177	0.005
p_1	1.235	0.003
p_2	-0.134	0.001
p_3	2.36×10^{-3}	2.7×10^{-4}
p_4	5.20×10^{-4}	6.1×10^{-5}
q_0	4.196	0.005
q_1	0.720	0.003
q_2	-0.0719	0.0011
q_3	-3.18×10^{-3}	2.6×10^{-4}
q_4	7.06×10^{-4}	6.0×10^{-5}
C_1	8.85	0.30
C_2	192 °C	6 °C

Table 4a. Certified Values of Viscosity and First Normal Stress Difference
with Expanded Combined Uncertainties at 0 °C

Temperature	Shear Rate	Certified Value of the Viscosity, η	Uncertainty in the Viscosity	Certified Value of the First Normal Stress Difference, N_1	Uncertainty in N_1
°C	s^{-1}	Pa·s	Pa·s	Pa	Pa
0.0	0.001000	383	12		
0.0	0.001585	380	11		
0.0	0.002512	382.4	9.9		
0.0	0.003981	382.9	9.5		
0.0	0.006310	384.0	9.2		
0.0	0.01000	383.1	9.0		
0.0	0.01585	382.9	8.8		
0.0	0.02512	379.8	8.6		
0.0	0.03981	375.1	8.3		
0.0	0.06310	365.5	7.9		
0.0	0.1000	350.0	7.3		
0.0	0.1585	328.4	6.6	16	11
0.0	0.2512	300.8	5.7	44	10
0.0	0.3981	268.5	4.8	92	11
0.0	0.6310	233.0	3.8	149	14
0.0	1.000	196.5	2.9	257	15
0.0	1.585	161.1	2.1	372	15
0.0	2.512	128.4	1.5	573	17
0.0	3.981	99.45	0.98	845	21
0.0	6.310	75.07	0.66	1219	29
0.0	10.00	55.59	0.44	1717	35
0.0	15.85	40.26	0.29	2363	44
0.0	25.12	28.67	0.20	3196	55
0.0	39.81	20.23	0.17	4251	69
0.0	63.10	13.88	0.15	5519	87
0.0	100.0	9.08	0.12	7126	107

Table 4b. Certified Values of Viscosity and First Normal Stress Difference
with Expanded Combined Uncertainties at 25 °C

Temperature	Shear Rate	Certified Value of the Viscosity, η	Uncertainty in the Viscosity	Certified Value of the First Normal Stress Difference, N_1	Uncertainty in N_1
°C	s^{-1}	Pa·s	Pa·s	Pa	Pa
25.0	0.001000	97.9	7.0		
25.0	0.001585	98.1	4.9		
25.0	0.002512	98.3	3.7		
25.0	0.003981	97.9	2.9		
25.0	0.006310	98.4	2.5		
25.0	0.01000	98.1	2.3		
25.0	0.01585	98.7	2.2		
25.0	0.02512	98.8	2.2		
25.0	0.03981	98.6	2.1		
25.0	0.06310	98.4	2.1		
25.0	0.1000	97.5	2.0		
25.0	0.1585	96.1	2.0	2.4	1.9
25.0	0.2512	93.7	1.9	5.5	1.8
25.0	0.3981	90.0	1.8	12.9	1.9
25.0	0.6310	84.6	1.6	26.5	2.1
25.0	1.000	77.6	1.4	50.1	2.5
25.0	1.585	69.2	1.2	87.6	3.3
25.0	2.512	59.98	0.94	148.2	4.6
25.0	3.981	50.56	0.72	236.8	7.4
25.0	6.310	41.44	0.54	377	12
25.0	10.00	33.04	0.39	585	16
25.0	15.85	25.60	0.28	880	21
25.0	25.12	19.36	0.20	1280	27
25.0	39.81	14.26	0.15	1800	36
25.0	63.10	10.22	0.14	2462	51
25.0	100.0	7.22	0.13	3319	63

Table 4c. Certified Values of Viscosity and First Normal Stress Difference
with Expanded Combined Uncertainties at 50 °C

Temperature	Shear Rate	Certified Value of the Viscosity, η	Uncertainty in the Viscosity	Certified Value of the First Normal Stress Difference, N_1	Uncertainty in N_1
°C	s^{-1}	Pa·s	Pa·s	Pa	Pa
50.0	0.001000	36.7	6.5		
50.0	0.001585	37.2	4.2		
50.0	0.002512	37.6	2.8		
50.0	0.003981	37.3	2.0		
50.0	0.006310	37.7	1.5		
50.0	0.01000	37.7	1.3		
50.0	0.01585	37.5	1.1		
50.0	0.02512	37.5	1.1		
50.0	0.03981	37.8	1.1		
50.0	0.06310	37.8	1.1		
50.0	0.1000	37.8	1.1		
50.0	0.1585	37.7	1.1		
50.0	0.2512	37.4	1.1		
50.0	0.3981	36.9	1.1	2.4	1.7
50.0	0.6310	36.1	1.1	5.1	1.7
50.0	1.000	34.8	1.0	10.7	1.8
50.0	1.585	32.79	0.99	21.5	2.0
50.0	2.512	30.20	0.93	41.0	2.7
50.0	3.981	27.05	0.84	68.8	4.3
50.0	6.310	23.54	0.74	114.5	7.6
50.0	10.00	19.95	0.64	203	11
50.0	15.85	16.41	0.53	345	16
50.0	25.12	13.14	0.43	558	23
50.0	39.81	10.24	0.34	860	33
50.0	63.10	7.75	0.26	1269	47
50.0	100.0	5.72	0.20	1797	67

Table 5a. Certified Values of the Storage Modulus G' and the Loss Modulus G"
with Expanded Combined Uncertainties at 0 °C

Temperature	Frequency of Oscillation	Certified Value of the Storage Modulus G'	Uncertainty in G'	Certified Value of the Loss Modulus G"	Uncertainty in G"
°C	rad/s	Pa	Pa	Pa	Pa
0.0	0.02512	0.605	0.058	7.98	0.15
0.0	0.03981	1.385	0.090	12.35	0.23
0.0	0.06310	2.97	0.14	18.85	0.35
0.0	0.1000	6.07	0.23	28.20	0.51
0.0	0.1585	11.70	0.37	41.11	0.73
0.0	0.2512	21.21	0.58	58.3	1.0
0.0	0.3981	36.21	0.88	79.8	1.4
0.0	0.6310	58.5	1.3	105.9	1.8
0.0	1.000	89.6	1.9	135.6	2.3
0.0	1.585	130.9	2.5	168.1	2.9
0.0	2.512	183.0	3.4	201.9	3.5
0.0	3.981	246.1	4.4	235.9	4.1
0.0	6.310	319.8	5.5	268.8	4.8
0.0	10.00	403.3	6.8	299.7	5.4
0.0	15.85	495.3	8.1	328.1	6.1
0.0	25.12	593.8	9.7	354.1	6.8
0.0	39.81	697	11	378.9	7.6
0.0	63.10	804	13	404.1	8.3
0.0	100.0	913	15	432.1	9.2

Table 5b. Certified Values of the Storage Modulus G' and the Loss Modulus G"
with Expanded Combined Uncertainties at 10 °C

Temperature	Frequency of Oscillation	Certified Value of the Storage Modulus G'	Uncertainty in G'	Certified Value of the Loss Modulus G"	Uncertainty in G"
°C	rad/s	Pa	Pa	Pa	Pa
10.0	0.03981	0.486	0.052	7.29	0.13
10.0	0.06310	1.099	0.082	11.35	0.21
10.0	0.1000	2.44	0.13	17.42	0.31
10.0	0.1585	5.08	0.20	26.29	0.47
10.0	0.2512	10.02	0.33	38.73	0.68
10.0	0.3981	18.53	0.52	55.45	0.95
10.0	0.6310	32.33	0.80	76.9	1.3
10.0	1.000	53.2	1.2	103.2	1.7
10.0	1.585	83.2	1.7	133.7	2.2
10.0	2.512	123.3	2.4	167.6	2.8
10.0	3.981	174.8	3.2	203.4	3.4
10.0	6.310	238.2	4.2	240.1	4.1
10.0	10.00	313.2	5.3	276.1	4.8
10.0	15.85	398.8	6.7	310.3	5.5
10.0	25.12	493.9	8.1	342.2	6.3
10.0	39.81	596.4	9.7	372.0	7.0
10.0	63.10	704	11	400.7	7.8
10.0	100.0	816	13	429.9	8.7

Table 5c. Certified Values of the Storage Modulus G' and the Loss Modulus G"
with Expanded Combined Uncertainties at 20 °C

Temperature	Frequency of Oscillation	Certified Value of the Storage Modulus G'	Uncertainty in G'	Certified Value of the Loss Modulus G"	Uncertainty in G"
°C	rad/s	Pa	Pa	Pa	Pa
20.0	0.03981	0.182	0.034	4.605	0.083
20.0	0.06310	0.430	0.053	7.26	0.13
20.0	0.1000	1.023	0.080	11.28	0.20
20.0	0.1585	2.27	0.13	17.38	0.30
20.0	0.2512	4.81	0.20	26.30	0.45
20.0	0.3981	9.57	0.32	38.91	0.66
20.0	0.6310	17.92	0.50	56.01	0.94
20.0	1.000	31.62	0.78	78.2	1.3
20.0	1.585	52.6	1.2	105.5	1.7
20.0	2.512	82.7	1.7	137.5	2.2
20.0	3.981	123.8	2.4	173.3	2.8
20.0	6.310	176.8	3.2	211.6	3.5
20.0	10.00	242.5	4.2	251.0	4.2
20.0	15.85	320.6	5.4	290.1	4.9
20.0	25.12	410.4	6.7	327.5	5.7
20.0	39.81	510.1	8.2	363.0	6.5
20.0	63.10	617.8	9.8	396.8	7.3
20.0	100.0	731	12	429.7	8.2

Table 5d. Certified Values of the Storage Modulus G' and the Loss Modulus G" with Expanded Combined Uncertainties at 30 °C

Temperature	Frequency of Oscillation	Certified Value of the Storage Modulus G'	Uncertainty in G'	Certified Value of the Loss Modulus G"	Uncertainty in G"
°C	rad/s	Pa	Pa	Pa	Pa
30.0	0.03981	0.077	0.023	3.028	0.055
30.0	0.06310	0.178	0.036	4.773	0.084
30.0	0.1000	0.440	0.054	7.48	0.13
30.0	0.1585	1.039	0.082	11.65	0.20
30.0	0.2512	2.32	0.13	17.94	0.31
30.0	0.3981	4.90	0.21	27.17	0.46
30.0	0.6310	9.78	0.33	40.19	0.68
30.0	1.000	18.38	0.52	57.92	0.96
30.0	1.585	32.46	0.80	80.9	1.3
30.0	2.512	54.0	1.2	109.2	1.8
30.0	3.981	85.2	1.7	142.6	2.3
30.0	6.310	127.7	2.5	180.0	2.9
30.0	10.00	182.6	3.3	220.0	3.6
30.0	15.85	250.7	4.4	261.3	4.3
30.0	25.12	331.7	5.6	302.3	5.1
30.0	39.81	424.5	7.0	342.1	5.9
30.0	63.10	527.3	8.6	380.4	6.8
30.0	100.0	638	10	417.2	7.6

Table 5e. Certified Values of the Storage Modulus G' and the Loss Modulus G"
with Expanded Combined Uncertainties at 40 °C

Temperature	Frequency of Oscillation	Certified Value of the Storage Modulus G'	Uncertainty in G'	Certified Value of the Loss Modulus G"	Uncertainty in G"
°C	rad/s	Pa	Pa	Pa	Pa
40.0	0.03981	0.030	0.019	2.047	0.039
40.0	0.06310	0.062	0.025	3.236	0.063
40.0	0.1000	0.200	0.038	5.101	0.096
40.0	0.1585	0.478	0.055	8.00	0.15
40.0	0.2512	1.142	0.088	12.45	0.23
40.0	0.3981	2.56	0.14	19.16	0.36
40.0	0.6310	5.39	0.22	28.94	0.54
40.0	1.000	10.69	0.36	42.76	0.80
40.0	1.585	19.97	0.57	61.4	1.1
40.0	2.512	35.03	0.89	85.5	1.6
40.0	3.981	58.0	1.4	115.1	2.1
40.0	6.310	91.0	2.0	149.8	2.8
40.0	10.00	135.7	2.8	188.5	3.5
40.0	15.85	193.2	3.9	229.7	4.3
40.0	25.12	264.1	5.1	272.2	5.1
40.0	39.81	347.7	6.6	314.6	6.0
40.0	63.10	443.1	8.3	356.1	6.9
40.0	100.0	548	10	396.3	7.9

Table 5f. Certified Values of the Storage Modulus G' and the Loss Modulus G" with Expanded Combined Uncertainties at 50 °C

Temperature	Frequency of Oscillation	Certified Value of the Storage Modulus G'	Uncertainty in G'	Certified Value of the Loss Modulus G"	Uncertainty in G"
°C	rad/s	Pa	Pa	Pa	Pa
50.0	0.03981	0.013	0.017	1.457	0.034
50.0	0.06310	0.034	0.021	2.316	0.058
50.0	0.1000	0.088	0.028	3.63	0.092
50.0	0.1585	0.241	0.041	5.73	0.15
50.0	0.2512	0.592	0.063	8.98	0.23
50.0	0.3981	1.37	0.10	13.95	0.37
50.0	0.6310	3.06	0.17	21.39	0.56
50.0	1.000	6.39	0.28	32.21	0.85
50.0	1.585	12.51	0.46	47.3	1.2
50.0	2.512	23.03	0.77	67.6	1.8
50.0	3.981	40.0	1.2	93.5	2.5
50.0	6.310	65.4	1.9	125.0	3.3
50.0	10.00	101.5	2.9	161.6	4.3
50.0	15.85	149.8	4.2	202.0	5.3
50.0	25.12	211.6	5.8	245.0	6.5
50.0	39.81	286.6	7.8	289.0	7.7
50.0	63.10	375	10	333.1	9.0
50.0	100.0	474	13	376	10

REFERENCES

[1] *Guide to the Expression of Uncertainty in Measurement*, ISBN 92-67-10188-9, 1st Ed. ISO, Geneva, Switzerland, (1993); see also Taylor, B.N. and Kuyatt, C.E., "Guidelines for Evaluating and Expressing the Uncertainty of NIST Measurement Results," NIST Technical Note 1297, U.S. Government Printing Office, Washington, DC, (1994); available at http://physics.nist.gov/Pubs/.

[2] Schultheisz, C.R. and Leigh, S.J., "Certification of the Rheological Behavior of SRM 2490, Polyisobutylene Dissolved in 2,6,10,14-Tetramethylpentadecane," NIST Special Publication 260-143, U.S. Department of Commerce, Technology Administration, National Institute of Standards and Technology, (2001).

[3] Macosko, C.W., *Rheology - Principles, Measurements and Applications*, Wiley-VCH, New York, (1994).

[4] Gordon, G.V. and Shaw, M.T., *Computer Programs for Rheologists*, Hanser Publishers, Munich, (1994).

Users of this SRM should ensure that the certificate in their possession is current. This can be accomplished by contacting the SRM Program at: telephone (301) 975-6776; fax (301) 926-4751; e-mail srminfo@nist.gov; or via the Internet http://www.nist.gov/srm.

www.ingramcontent.com/pod-product-compliance
Lightning Source LLC
Chambersburg PA
CBHW081830170526
45167CB00007B/2772